Coal-Mining Safety in the Progressive Period

Published for
the Organization of
American Historians

COAL-MINING SAFETY IN THE PROGRESSIVE PERIOD

The Political Economy of Reform

William Graebner

The University Press of Kentucky

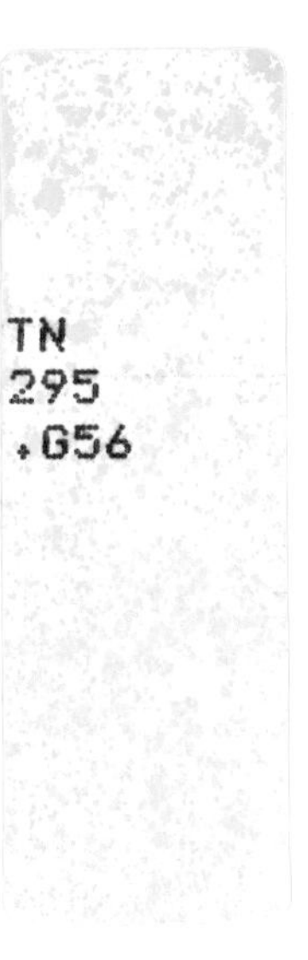

ISBN: 0-8131-1339-3

Library of Congress Catalog Card Number: 75-38215

A statewide cooperative scholarly publishing agency serving Berea College, Centre College of Kentucky, Eastern Kentucky University, Georgetown College, Kentucky Historical Society, Kentucky State University, Morehead State University, Murray State University, Northern Kentucky State College, Transylvania University, University of Kentucky, University of Louisville, and Western Kentucky University.

Editorial and Sales Offices: Lexington, Kentucky 40506

TO DIANNE

Contents

List of Tables

Preface

A FEW YEARS AGO, when I first began to put these materials together, I was convinced that I had come across another Progressive triumph. At the center of that triumph I placed the United States Bureau of Mines, an agency admittedly lacking coercive powers but nonetheless embodying a fresh, national, scientific, ongoing approach to the serious problems of mine safety. There is still some of that emphasis in this book, but I am now less convinced of the propriety of using the bureau, the Progressive period's unique solution to mine accidents and explosions, to exemplify a thoroughly modern institutional response to industrial dislocation. My skepticism derives in part from an examination of mine safety in the states, perhaps the major theater of activity, where the same coal-mine operators who had pressed for a national bureau resisted their state legislatures and inspection services and finally, in their misplaced enthusiasm for uniform state legislation, opposed meaningful national reform. In short, if the coal-mining safety movement may be considered exemplary, then the Progressive period was characterized not by centralization and organization, but by the persistence of decentralization and disorganization. At the heart of this distended society was the Progressive political economy: a hydra of political federalism and competitive capitalism, each a form of decentralization, each feeding on the other, together responsible both for the tragedies in the mines and for the failure to find adequate solutions.

This book is a blend of several approaches to history and varieties of historical writing. I have attempted to place the subject of coal-mining safety in differing perspectives. Although the study focuses on the process of reform in a federal system, I have devoted some attention to technology, bureaucracy, and those miner and operator attitudes which lie outside a traditional political scheme. In general, I have rejected narrative chronological history in favor of organization around subject areas and occupational groups. In the final chapter, I have tried to make my interpretative frameworks as obvious and explicit as possible.

I am grateful to the State University of New York Research Foundation and to the Fredonia Foundation for summer grants in support of this study and to the helpful staffs of research libraries across the country. A special thanks goes to Stanley Brown of the Federal Records Center at Suitland, Maryland, whose cooperation saved me many hours and frustrations. Mary Notaro, who typed the final manuscript, has my appreciation for her precision and tolerance of my errors. The dissertation from which this book has grown was written at the University of Illinois, Urbana, under Clark Spence, and was read by Frederic Jaher and Thomas Krueger, who served on my dissertation committee. I have profited considerably from their observations and suggestions. Robert W. Harbeson of the Department of Economics at Illinois had a clear though indirect impact on this work through his course in competition and monopoly. K. Austin Kerr of Ohio State University read the completed study in full. I am thankful for his expertise in the history of mine safety and for a thoughtful overall critique. Marilee Sargent and Jerry Clore, friends since 1968, have never read this manuscript, but their influence upon it, and upon my life and world view, has been enormous. Dianne Bennett, my wife, has unselfishly given hundreds of hours and her talents in researching, writing, and editing to this project. She has spent her vacations in manuscript collections, her summers near research libraries. I often wonder, had our roles been reversed, if I would have been as willing to donate *my* time and energies. Her direct contributions to my work have decreased in recent years, and with this I am equally pleased, since it indicates that Dianne has found her own sources of fulfillment and happiness.

Introduction

IN NOVEMBER 1968 an explosion in the "safe" Number 9 mine of the Consolidation Coal Company at Farmington, West Virginia, killed seventy-eight miners and produced a brief but potent public outcry for new federal safety legislation. The prototype for that scenario of disaster and reform belongs to the Progressive period. It began in December 1907, ten miles from Farmington at Monongah, West Virginia, where 361 miners died in an explosion in the "safe" Monongah 6 and 8 mines of the Fairmont Coal Company.[1] Monongah, and a number of other major disasters occurring in the same month, put the coal-mining safety problem in proper perspective; it made it a public disgrace. It proved an important stimulus to reform in the states, where a limited coal-mining safety movement had been instrumental in fostering new legislation since 1905. More important, Monongah made national reform likely, for it now became possible for Joseph A. Holmes, chief of the Technologic Branch of the Geological Survey and protagonist in the extended drama to follow, to obtain a hearing for his idea of a national educational and scientific agency to deal with the problem of mine safety. By 1910 a coalition of interests, including miners, inspectors, coal and metal mine operators, conservationists, and bureaucrats, had created the United States Bureau of Mines.

In one sense, the bureau was the cautious first step in the national politics of mine safety; it would be followed in the next few years by a major campaign for uniform state legislation (a form of nationalization) and in the 1940s and 1950s by federal safety standards and inspection. In another sense, the bureau was the end product of rapid changes in the coal industry; the miners at Monongah were victims of economic growth, increasing competition, and changing technology. Production of bituminous coal approximately doubled every ten years beginning in 1840, with increases more gradual (excepting the war years) after 1910. Although the period from 1890 to 1920 was one of steady increase in demand for coal, competition, much of it interregional, also grew in intensity. By 1900 more than twenty-five states mined coal, and Illinois, Pennsylvania, Indiana, Ohio, and West Virginia had emerged as the industry's major producers in the United States. Attempts to satisfy increasing industrial demands for coal created new mine-safety problems and exacerbated old ones. As the mines went deeper, they exposed potentially dangerous concentrations of explosive methane gas; as they became more mechanized (machines cut about 25 percent of the nation's coal in 1900 and 42 percent in 1910), coal dust developed as a major health and safety hazard.[2] Among the new hazards were electricity, now beginning to be widely deployed in coal mines; a system of payment for miners—mine-run—which encouraged misuse of explosives since, under it, miners were paid for all the coal brought out, rather than for only the larger pieces; and the new immigrants, usually lacking in mining experience and English-language capabilities.[3]

The matrix of institutions and processes which the Progressive period brought to bear on mine safety confronted an exceedingly critical situation, but it was one for which there was substantial precedent and preparation. Accidents were reported in the United States as early as 1825, with the first near-disaster in 1856 when four men were buried alive but ultimately rescued. Not until the Avondale disaster of 1869 in the Pennsylvania anthracite territory did coal-mining safety attract sufficient attention to evoke legislative action, although miners had been attempting to secure safety legislation since before the Civil War.[4] The states legislated on numerous facets of coal-mining safety after 1870, paying particular attention to ventilation and systems of mine inspection. All the legislation proved inadequate. Major disasters rocked Braidwood, Illinois, in 1883; Pocahontas, Virginia, in 1884; and the Frick Coal and Coke Company, Pittsburgh, in 1891. By 1890 fourteen states had experienced at least one major explosion.[5] A miner, recalling his youth in the coalfields of West Virginia, described the mood surrounding these tragedies:

> Not a week passed but what tragedy touched some home. When a housewife, intent upon her duties, chanced to glance through the window and see a group of miners bearing an improvised stretcher between them, she spread the alarm. In a twinkling women were on the porches, wiping hands on aprons, calling to one another. And before the grim-faced bearers of the groaning burden, carbide lamps still shooting flame from their foreheads, were halfway to the doctor's office, a multitude of folk trailed in their wake. Anxious, distrait women and children, uncertain of the fate of their loved ones, demanded to know the identity of the victim. When the dreaded news was revealed, the women gathered around their hysterical sister and offered com-

fort. Presently they returned to their household duties, thankful that a merciful God had seen fit to once again spare husband, or son, or brother.[6]

The new century began inauspiciously with the Winter Quarters mine disaster in Schofield, Utah, which killed 201 persons. In 1906 seventeen major disasters killed a total of 235 miners; in 1907, eighteen killed 918 miners; in 1908, eleven killed 348; in 1909, nineteen killed 498; and in 1910, when Congress created the bureau, nineteen major disasters killed 485 miners. The problem of industrial safety—in the nineteenth century a malingering illness to which occasional relief was administered—had by the twentieth century become an epidemic of crisis proportions, demanding immediate attention.[7]

During the Progressive years, the coal industry's particular version of this industrial crisis was approached on four distinct yet converging planes. Of the four, the most traditional and the least productive plane encompassed private activities related to mine safety and the attitudes paralleling or underlying those activities. Some coal-mining companies, usually the largest ones, undertook to improve conditions in their mines, installing safety equipment and training safety foremen and rescue personnel, even when not required by law to do so. They pursued safety activities through a variety of organizations, including local, state, and regional coal-operator associations, the American Mine Safety Association, and the National Safety Council. Yet operators more commonly were circumspect of even the innocuous activities of these organizations, and if the reports of the state mine inspectors are reliable, a majority failed to keep their mines in minimally satisfactory condition, frustrated enforcement of safety codes, and cooperated only under duress. Miners, for their part, brought to their work a set of safety attitudes and work habits which contributed to mine-safety problems. They often were guilty of inadequately timbering their working places, of failing to undercut coal before shooting, and of riding illegally on mine cars. Many, particularly the recent immigrants, were ignorant of elementary mining techniques and unable to read mine-safety materials and signs or to understand oral instructions from working partners and supervisors. The claim of miner carelessness, although politicized to the point where it virtually became a rationale for the failures of the entire mine-safety movement, appears to have had some basis in fact. In his organized role, as part of the United Mine Workers of America (UMWA), the miner played a more positive yet still ambivalent part in the safety movement. The union was the primary advocate of new state mine-safety legislation and, unlike its constituency, an active proponent of enforcement through inspectors and the contract grievance procedure. The organization also participated, though less zealously, in the campaigns to create and fund the Bureau of Mines. Nonetheless, the UMWA remained essentially committed to the goals of organization and membership, and with the possible exception of the years 1905–1910, safety remained a decidedly peripheral concern. A third group, the state mine inspectors, showed more consistent interest in the cause of safety. Most inspectors escaped dominance by capital or labor, worked to enforce and improve the laws, and, with the operators, were the major advocates of uniform state legislation.

The three other planes of activity were political, each associated with a level of governmental activity—national, state, and interstate. At the national level, the primary political response of the Progressive period to coal-mining safety was the Bureau of Mines, a research- and education-oriented organization totally lacking in coercive capability. The bureau was in part a product of a campaign for national action which had its origin within the federal bureaucracy, in the work of Joseph A. Holmes. That campaign became public following an incredible series of mine disasters in December 1907, and it was thereafter shaped and dominated by an active group of coal operators led by West Virginia's A. B. Fleming. Although the coal-mining safety issue was the major catalyst in the bureau's creation, there is some doubt whether it would have been established without the aid of western metal mine operators and their cooperation with coal interests within the American Mining Congress. Placed beside the consistent, aggressive, and well-organized political tactics of the business community, the contributions of scientists, engineers, inspectors, the UMWA, and a variety of humanitarian progressives and socialists seem sporadic and ineffectual. Once the Bureau of Mines had been established, operators shifted their attention to support of Holmes as director, and here, too, their efforts were eventually rewarded, though only after a protracted struggle with federal bureaucrats who were opposing Holmes and pressing their own candidate.

Federal initiatives in mine safety, first under the United States Geological Survey, then under the Bureau of Mines, reflected the political origins of the mine-safety issue. The influence of the mine disaster, especially Monongah in 1907 and the Cherry, Illinois, mine fire in 1909, was unmistakable in the policies and emphases of both agencies. The Geological Survey sought to isolate the cause of the mine explosion and to counter the opposition of miners and operators to a theory of explosions based on coal-dust particles, and the bureau devoted an increasing proportion of its coal-mining safety energies to its popular rescue program using stations and movable cars. The bureau's conservative political origins and consequent lack of coercive authority also were apparent in its reliance on a vigorous public relations program, aimed at miners, operators, and public officials, and in its search for productive cooperative relationships with state agencies and private organizations like the National Fire Protection Association. But for coal-mining safety, perhaps the most significant aspect of the bureau's first decade was its gradual shift away from the coal industry in response to the political demands of western metal mine operators and away from safety as public interest in matters of safety diminished.

Within individual states, the politics of mine safety followed an expected pattern: mine workers, usually backed by inspectors, consistently advocated additions to and modifications of state legislation, while operators presented the greatest obstacle to reform. Under the consistent pressure of explosions and rising death and accident rates, state political systems responded with new legislation, most of it passed between 1905 and 1915. In several states, participants in the political process attempted with some success to replace confrontation with cooperation as a decision-making method, developing commissions of operators, miners, and interested public officials which worked to compromise differences and present lawmakers with legislation agreeable to all parties. This creative mechanism and the legislation it produced were, however, at the mercy of the enforcement process. Inspection staffs were consistently overworked, often of low quality, and occasionally politicized beyond objectivity; and they seldom had the full cooperation of other elements in the safety equation. Miners, operators, mine officials, and coroners' juries all manifested a lack of commitment to enforcement.

The politics of coal-mining safety took place upon still a third political plane, this one interstate. Unwilling to countenance national safety legislation but aware of the inadequacies of a purely state system, coal-mine operators in the American Mining Congress and inspectors in the Mine Inspectors' Institute of America (founded in 1908) led a major industry movement toward uniform state legislation. This administrative device for equalizing the costs of safety was never effected, largely because operators and inspectors alike were unable to resolve their internal differences.

These public and private solutions brought to bear on the safety problem in the Progressive years proved to be no panacea, as Farmington attests. In fact, the most recent study of coal-mine accidents concludes that "for the first quarter of the century it is difficult to observe any clearly marked course of improvement in the fatal accident rates of American coal mines . . . after the turn of the century little sustained improvement is visible."[8] The first two bureau chiefs, Joseph A. Holmes and Van H. Manning, proffered more favorable evaluations. In early 1913 Holmes noted a "gratifying improvement in the situation"; six years later Manning claimed that the coal-mining safety movement had saved some 5,000 lives in less than ten years.[9] Acknowledging the inadequacies of any statistical method and the inaccuracies of all compilations for the 1900 to 1920 period, statistics indicate some improvement in the overall coal-mining safety situation after 1906. From that date until 1920, a period in which bituminous coal production increased by almost 66 percent, major disasters and the fatalities they produced declined.[10] Between 1906 and 1910 there were 84 major disasters and 2,492 resulting fatalities; from 1911 to 1915, 59 major disasters and 1,718 fatalities; from 1916 to 1920, 45 major disasters and 773 fatalities.[11] Thus the number of major disasters was reduced by almost half in the last five-year period as compared with the first, and the total number of fatalities in major disasters in 1916–1920 was less than one-third of the 1906–1910 total. Although yearly fatality totals (i.e., including those outside of major disasters) remained virtually the same over the fifteen-year period in question, the fatalities per million tons of coal produced decreased substantially. Figure 1 shows graphically the relationship between fatalities and production of coal from 1906 to 1920.[12]

The coal-mining safety movement, therefore, had at least witnessed a reduction in the relative number of fatalities and a substantial decrease in the absolute number of major disasters and fatalities from these disasters. The failure or weakness of the movement in these years, however, is implicit in these claims, for coal-mining safety was explosion-, disaster-, and fatality-conscious. While progress was substantial in those areas related to disasters—explosives and gas and dust explosions—the much more pervasive although less dramatic problem of fatalities from falls of roof and coal and from the hazards of haulage and electricity was almost completely disregarded. As Table 1 indicates, more persons were killed in 1920 than in 1907 by falls of roof and coal, by haulage accidents, and by electricity.[13] Between 1906 and 1910, gas and dust explosions accounted for only

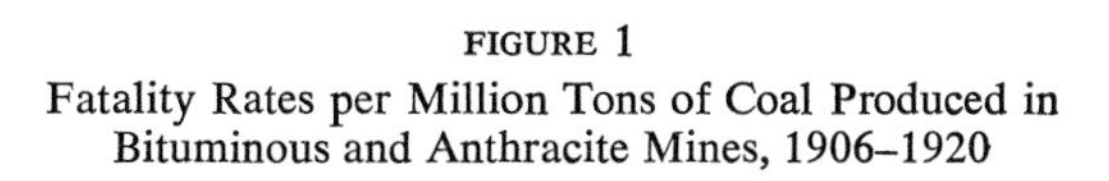

FIGURE 1

Fatality Rates per Million Tons of Coal Produced in Bituminous and Anthracite Mines, 1906–1920

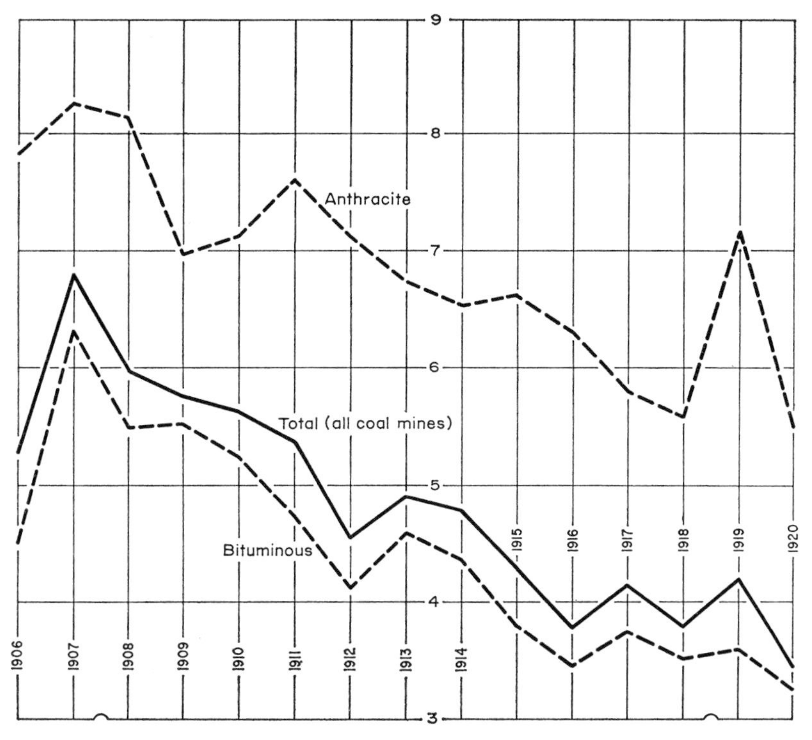

18 percent of the fatalities in the coal mines, yet received much of the attention of safety experts; falls of roof and coal accounted for 44 percent but were not studied, even after 1910. Moreover, although there were ten to fifteen nonfatal accidents for each fatal one, injuries aroused little public interest and thus were not the subject of serious professional concern.[14]

Though the efforts of safety experts were somewhat misplaced and bureau efforts misguided, the essential sources of frustration lay elsewhere. As an industrial problem and as the subject of reform efforts, coal-mining safety was deeply influenced by two systems operative at the turn of the century: economic competition and federalism. Each system was harmful enough by itself; together, they formed the nexus of a political economy which was not only responsible for many of the coal industry's safety problems but which also prevented or militated against the solutions of the Progressive period. These frame-

TABLE 1
Underground Fatalities at Bituminous Coal and Lignite Mines, by Principal Causes, 1906–1920

Year	Falls of roof and coal	Haulage	Gas or dust explosions	Explosives	Electricity
1906	826	192	219	115	52
1907	911	260	911	134	49
1908	829	229	320	109	53
1909	975	240	264	122	52
1910	1,061	295	477	113	76
1911	1,007	294	331	72	92
1912	1,043	296	255	70	75
1913	1,007	344	464	63	79
1914	903	309	305	56	86
1915	818	269	270	76	85
1916	850	330	183	60	81
1917	965	433	319	55	76
1918	1,052	446	103	85	77
1919	878	310	149	57	64
1920	937	342	124	82	70

works have little relevance to the anthracite industry, which largely explains why this study draws almost entirely on bituminous coal. The coals also had somewhat different physical properties. Because it burned cleanly, slowly, and steadily, anthracite coal was used primarily as a household fuel. Bituminous coal, or soft coal, has had primarily industrial uses. Anthracite mines were usually deeper and more gaseous than bituminous mines in this period, and their thick, steeply pitching veins made propping more difficult than in bituminous mines. In general, more blasting was necessary to bring down the coal. On the other hand, anthracite carbon content was so high that its mines were not subject to coal-dust explosions. The safety problems in a number of areas, including haulage, fires, electricity, and explosives handling, were similar.

From 1890 to the First World War, the coal industry resisted the considerable efforts of its operators to bring order to what was widely regarded, by government officials and mine workers as well as operators, as an excessively competitive industry. In most major coal markets east of the Mississippi, coals from several states were in competition. Pennsylvania, West Virginia, and Ohio coals moved by rail and lake steamer to the distribution and marketing centers of Chicago, Milwaukee, and Duluth-Superior, where they were competitively priced with lower-grade coals produced in Illinois and Indiana. Kentucky coals shipped north significantly influenced the market in every midwestern state from Ohio to Minnesota. The nation's two largest producers after 1910, Pennsylvania and West Virginia, effectively dominated the trade in the eastern tidewater, in select western markets like Cincinnati, and in one particular product market, coke. Seldom did Illinois and Indiana coals enter these or other eastern markets. This interstate competition reflected the wide availability of coal and the labor to mine it, the relatively low capital requirements for its development, and the ease with which one coal (or an alternative energy source) could be substituted for another. The result was a low-profit industry with chronic and growing excess capacity and an extremely low level of concentration. The bituminous coal industry had its large firms, but none had significant market power beyond the local marketing area.[15] These conditions prompted this remark from operator attorney D. W. Kuhn: "Among the Falstaff army of industries of this country, too poor to fight, too cowardly or too virtuous to steal, the coal mining industry presents itself as one of the most bedraggled members of these ragged recruits."[16] The military analogy also seemed appropriate to Pennsylvania's chief mine inspector, James Roderick. "The rapid growth of the industry," he said, "has prevented systematic development and today the operators constitute a great army of antagonistic elements and unorganized forces . . . they continue to indulge in a cut-throat war-fare."[17]

Paralleling this economic configuration was a political system in which social legislation—and that included safety legislation—was conceived to be the proper and virtually exclusive province of the states. This was not the view of a few strict constructionists or of conservative business reformers. Of the many proposals for some kind of national action on mine safety, few would have invested the national government with powers of inspection and enforcement, and they were carefully avoided by Congress and, with the exception of the UMWA, by affected interest groups. A national agency with coercive capability, a national inspection force, national mine-safety laws—all these were believed, by reformers and their opponents alike, to be outside the normal range of political options. Federalism and the resistance to national legislation which it implied functioned on several levels. It was an article of faith, a basic value, a belief so common and so widely accepted as to go unquestioned. It was also part ideology, in the sense that people continued to hold it dear long after it had proved to be of declining usefulness. Finally, federalism had become a functional element in the economics of competition. As coal-mine operators competed for access to markets and for adequate supplies of railroad cars, they were also forced to compete in their state legislatures by opposing safety and health measures which would add significantly to their production costs.

The economics of competition was responsible for the poor safety performance of the coal industry in a variety of other ways. It insured the existence of numerous small firms which would resist enforcement of state codes and which could ill afford the latest in safety devices or who (and this is functionally identical) believed they could not afford them. United States Steel's substantial reputation in safety was achieved in large, captive mines, isolated from the coal industry's competitive markets. Competitive conditions also encouraged division within the industry, destroying potentially helpful trade associations and seriously impeding the uniformity campaign. As coal operators sought to cut costs in this labor-intensive industry, they brought in cheap labor and lowered wages. The first action reduced the skill level of the labor force; the second placed considerable pressure on the miner to rush the mining process in order to earn a living wage. To the extent that the stereotype has some validity, the careless miner was less the product of pride, independence, and stubborn individualism than the victim of forces generated by the competitive structure of the coal industry. Competition also had a substantial impact on union objectives, since the logical UMWA response to a decentralized economic framework was its emphasis on mining rates.

Together, economic and political decentralization produced the ineffectual politics of coal-mining safety. In the states, operators, even under great public pressure, resisted legislation which would place their states (and firms) at a competitive disadvantage. Aware that demands for reform in the states could not be effectively resisted, operators attempted to transcend this menacing situation and to shift the locus of reform activity into national and interstate politics. This effort, which in its broad outlines was approved by every coal-mining interest group, centered around the Bureau of Mines and uniform state legislation. The bureau was in part an attempt to arrive, through science, at a body of reasonable and widely acceptable knowledge which, if applied uniformly in the states, would at least affect the entire industry and prejudice no single group of operators. The movement for uniformity was a more direct attempt to equalize the burdens of safety legislation, to achieve the benefits of a centralized politics without centralizing the political system. This idealistic solution was at once the great hope and the great failure of the Progressive coal-mining safety movement.

CHAPTER 1

Business, Bureaucracy, and National Reform

CONCERN FOR the hazards of coal mining far antedates the Progressive period, but reform interest in coal-mining safety before 1900 was limited almost entirely to the states, finding an outlet in state (and occasionally county) mine-safety legislation. The states continued to legislate on the subject after 1900, but the new context of the legislation and some procedural innovation did not mask the fundamentally traditional quality of state politics and political alignments. What was new from 1900 to 1920 was the involvement of the federal government in coal-mining safety through the United States Geological Survey and the United States Bureau of Mines. This admittedly circumscribed and limited federal interest was the product of several historical factors, usually operating concurrently but to some extent chronologically separable. Before December 1907 only a few persons in the federal bureaucracy showed interest in mine safety outside of the federal territories, but that bureaucratic interest was sufficient to validate the claim that the reform movement began within the bureaucracy.[1] A series of explosions of monumental proportions occurring in December 1907 brought the problem of coal-mining safety national attention, producing an urgent demand for solutions to the scientific problems basic to coal-mining hazards. At this point the nation's press —newspapers and magazines—demanded national action, and the issue very rapidly became politically popular, making it difficult for politicians to oppose rigidly new legislation. At the same time, groups with vested interests in mine safety or in the political accoutrements of national mine safety (i.e., a bureau, research facilities) added their support to the movement. Miners, coal and metal mine operators, inspectors, conservationists, scientists—each group contributed something to the political compromise which finally produced the Bureau of Mines. Of these groups, the mine operators exerted the most influence within the political process, and the bureau, in spite of its origins and nurture in the federal bureaucracy, can properly be considered a business reform.

Federal lawmakers were aware of mine-safety problems as early as 1885, when Henry George, testifying before the Senate, said of coal miners, "They are constantly in danger; never out of danger; they do not know at what time a piece of 'horse flag' may fall and crush them to death." In general, however, this investigation of the relationships between capital and labor, insofar as it dealt with coal mining, emphasized the methods by which the miners were paid and their living conditions.[2] Another investigation into capital and labor in the mining industry, made in 1900 by the United States Industrial Commission, solicited testimony from operators and miners on all aspects of the industry, but references to safety were not many and tended to deemphasize the need for legislation. One operator, for example, testified that he did not know "any class of labor that is better protected than that engaged in the mining industry, so far as legislation is concerned."[3] The most consistent call of those testifying was for uniform mine-safety legislation—similar laws in all major coal-mining states. David Ross, secretary of the Illinois Bureau of Labor Statistics, said operators and mine workers alike would benefit from uniform mine legislation. George W. Schluederberg, an operator from Pittsburgh, supported uniform legislation on the grounds that it would place all operators on a competitive basis. John Mitchell, president of the United Mine Workers, favored uniform legislation as a way of strengthening laws in states where he saw them as inadequate. This general demand for uniform mine legislation was noted in the review of evidence by the commission, but the commissioners' report also stated that the safety laws of Illinois, Indiana, Ohio, and Pennsylvania were said to be good and inspection thorough.[4]

There were nineteenth-century precedents for national safety legislation. After 1852 the federal government operated a steamboat-inspection service at an annual cost by 1900 of almost $500,000; Congress appropriated funds for lighthouse stations and for rescue work incident to these stations; and after 1908, federal law required safety appliances on common carriers.[5] The national government became directly involved in legislating for mines in 1891 with the enactment of safety legislation for mines in the territories.[6] In 1900 the 1891 act was amended to include sections requiring, when practicable, the watering down of coal dust or the removal of the dust if water were not available and shot-firing by designated miners with the men out of the mine. The law as it was signed by the president on June 30, 1902, applied only to mines in the Indian Territory employing twenty or more miners. The legislative process here and the reports appended to the bill show that at least some members of the national legislature had knowledge not only of the general safety problem in coal mines but also of specific safety problems—ventilation and coal dust.[7]

These federal reports and legislation were obvious precedents for federal involvement in the mining-safety area, but the years from 1900 to 1907 witnessed no public outcry for safer mines. Indicative of press indifference toward mining deaths is a telephone conversation between a reporter and a coroner's deputy overheard by Crystal Eastman while she was doing research for the *Pittsburgh Survey* in 1907:

> Reporter: [ ].
> Coroner's Deputy: "No, we haven't got anything for you today, Jim. —Well,—hold on.—There's a man killed by a fall of slate out at Thom's Run. You don't want that, do you?"
> Reporter: [ ].
> Coroner's Deputy: "That's what I thought. No, there ain't anything else. So long."[8]

The *United Mine Workers Journal* commented on the problem of public apathy in 1900:

> One astounding and alarming feature in connection with the increased number of mine catastrophes in this country is the apparent indifference upon the part of those directly affected and the public in general, their apathy seeming resignation and tendency to look upon such terrible cyclones of death as coming in the regular course of human events, being purely accidental, and, therefore, not preventable. The daily press simply records the event in a matter-of-fact manner, and, after expressing the usual formal sympathy, dismisses the subject entirely without any appeal for better conditions, an investigation or other manifestation of interest, and the public . . . are disposed to look upon this as a visitation from God upon those whose lot it is to enter the dark and gloomy caverns of the earth in search of her treasures, and the whole affair is dropped from view.[9]

Writing in the *Century Magazine*, Jay Hambridge commented that even in the mining communities, "A death by violence is noted to-day, but to-morrow it is a fact remote, and is recalled by association of idea with some other incident."[10]

The *Journal*, the official organ of the United Mine Workers of America, was an exception to the general apathy toward mine accidents, but even the *Journal*'s interest in the question was subdued until 1905, when a large Alabama disaster early in the year brought to its pages an increased sense of urgency and a flood of suggestions, letters, studies, editorials, and legislative news dealing with coal-mining safety. Although the daily press reported coal-mining accidents, before 1907 neither it nor the national magazines took an editorial position on the subject or suggested federal action as a solution to the problem. There were several possible reasons for this inaction. First and most important was the nature of death in the industry, with most deaths occurring from causes other than explosions, and about half resulting from unnewsworthy falls of roof and coal which killed one or two miners at a time. In 1906, for example, the year before coal-mining accidents became a national political issue, there were 1,504 underground fatalities in bituminous coal and lignite mines, but only 219 of these were attributable to explosions. And of the seventeen official disasters (five or more deaths) in that year, none resulted in more than thirty-five deaths.[11] In 1906 a coal-mine explosion resulting in thirty-five deaths was news, but its shock value was rapidly diminishing and it could hardly be expected to produce a call for action on the national political scene.

Second, coal mining was not easily recognizable as the most dangerous of occupations, and other accident-prone occupations vied with coal mining for attention. Interstate railroading was a natural subject for national safety legislation. Railroad accidents were common, visually spectacular, and they directly affected the middle class which traveled by train. Hence, railroad rather than mine accidents received most of the national publicity. *Collier's*, for example, ran a two-page picture spread dealing with twenty-four railroad accidents which had killed 188 persons in a thirty-day period in 1907.[12] Of all the muckraking magazines, B. O. Flower's *Arena* may have been the most effective in dealing with railroads and the issue of regulation, but in all of 1907 and 1908, the *Arena* ran only one article on industrial accidents of any kind, and that one was on railroads. Other national magazines, such as *World's Work*, the Republican *Outlook*, and *Leslie's Illustrated Weekly*, dealt with the railroad issue, attempting to find the element responsible for industrial accidents. *Review of Reviews*, *Scientific Monthly*, and *Nation*, none of which were muckraking journals, also looked at the problem.[13] In general, this magazine coverage of industrial accidents contributed to an accident-prevention climate and put the problem of industrial accidents squarely before the public.

On December 6, 1907, a blown-out shot touched off the most disastrous mine explosion in the history of mining in the Western Hemisphere. Three hundred and sixty-one men lost their lives that day in Monongah mines 6 and 8 of the Fairmont Coal Company, Fairmont, West Virginia. Thirteen days later, 239 died at the Darr mine in Jacobs Creek, Pennsylvania.

One opinion of the cause of the Monongah explosion was voiced by James Sinnot of Chatham, Illinois, in a song entitled "The Monongah Disaster":

Oh, Monongah! Oh, Monongah!
Where 400 lives were sacrificed
By your gassy mines blown up.
Someone has been neglectful
Or in their duty sure did fail
And thus we read of neglect, indeed,
At those mines of Monongah.

Let us ask of West Virginia,
Is it right to kill wholesale?
And if she has got mining laws,
If so, why did they fail?
If your laws are not good,
Have them repealed, you should.
Make ones that will not fail;
Don't encore, we do implore,
Such scenes as Monongah.

To read of such a horror,
Forget we never will.
You have got no right from God or man
Thus human beings to kill.
It shows you've been neglectful
For your miners' lives to care,
And thus we read of graft and greed
At those mines of Monongah.

A committee of the West Virginia legislature, established earlier in the year in response to previous explosions, disagreed with Sinnot, concluding from its testimony that the Monongah mines were well equipped and modern, and that the company controlling the Monongah mines had an almost unsurpassed reputation for safety.[15] R. D. Nuzum wrote to A. B. Fleming, former governor of West Virginia and owner of the Fairmont Coal Company: "I can hardly believe that such a disaster could come upon a mine so well looked after as are the mines of the Fairmont Coal Company, but am forced at the last to credit the account."[16] The *United Mine Workers Journal* challenged this opinion, charging that faulty mine inspection was responsible for the explosion. The Monongah mines, said the *Journal*, were operated without two openings, contrary to West Virginia law, and neither in those mines nor in the Naomi mine was ventilation consistent with state requirements. "Chief Mine Inspector (J. W.) Paul of West Virginia," continued the *Journal*, "has held the office for nearly two decades. . . . His administration of his office is marked by one long bloody trail of human slaughter, caused by negligence, inefficiency, by wanton nullification of every mining law in the state."[17]

This opinion of the *Journal* aside, many thought the Monongah mines to be well equipped for safety and concluded that if the Monongah mines blew up, so could any mine. Moreover, explosions had occurred and were to occur in other mines with reputations for safety, with a cumulative effect significant for mine-safety reform. If the safest mines could explode, then the normal channels of accident prevention, state legislation and inspection, were simply insufficient. The West Virginia committee report said this explosion of a reputedly safe mine was illustrative of the limits of scientific knowledge: "It was conceded that this explosion was a dust explosion, yet no one has been able to give all the elements that may be present to create a dust explosion. This reference to the Monongah mines and to the conditions that were thought to exist there at that time are especially made for the purpose of impressing upon the legislature the fact that the legislature cannot reach the cause or provide a remedy which will prevent future explosions, until that cause is known. This must be the result of future study and experiment."[18] The Monongah disaster and others, because they occurred in mining properties considered safe, cast doubt on the safety of all coal mines and therefore were a direct stimulus to federal scientific research in coal-mining safety.

The national magazines were surprisingly slow to treat the mine explosions editorially. *Leslie's Weekly*, which according to historian Louis Filler gave the "accident issue enough momentum and drama to make it national," had nothing on its December 19, 1907, editorial page and carried only a picture story of the Monongah disaster in its

December 26 issue. Again on January 2, 1908, *Leslie's* carried a picture story on coal-mine explosions, but it had no suggestions for solving the problem. *Collier's* reported the "wholesale slaughters in coal mines" but launched no campaign for reform and did not mention coal-mining safety again for some time. It remained for Edgar Allen Forbes of *World's Work* to begin suggesting in the national press some answers to the coal-mining safety problem. In the February 1908 issue in an article entitled "The Human Toll of the Coal Pit," Forbes presented his thesis that the causes of accidents were similar worldwide and thus what had worked in Europe to prevent coal-mining accidents would work here. Forbes listed eight specific techniques designed to reduce accidents (including, for example, the use of shot-firers); in the process he attacked miner ignorance and inefficient inspection, "the weakest link in the chain." Forbes concluded that the best solution was the proposed national bureau of mines, which would function as a clearinghouse for information.[19]

In contrast to the national magazines, the nation's newspapers responded immediately with demands for federal investigative action. The *Pittsburgh Chronicle*, in a December 23, 1907, editorial, called for investigation by government commission. The *Washington* (D.C.) *Star*, while noting that the federal government had no direct jurisdiction over the matter, since the mines were exclusively under state control, emphasized the necessity of an immediate and thorough probing of the causes and circumstances of recent catastrophes. The *Detroit Free Press* pointed out the need to exploit European expertise in gas, dust, and explosives. The *Herald-Transcript* of Peoria, Illinois, succinctly stated in its December 23 issue: "If the Federal Government has any jurisdiction in the matter it should lose no time in acting."[20]

There is not much evidence here to assign the press a major historical role in the coal-mining safety movement. The national press neither created the issue of coal-mining safety nor uncovered previously unknown information about coal-mining accidents. It did not present solutions until the accident situation was so critical that editorial comment could not be avoided. Muckraking in the sense of interpretive and investigative reporting simply was not present in the coalfields. The press would have its influence on public opinion, of course, but fundamentally the press, like the general public, was the follower of events.

The government bureaucracy played a similarly limited but more dynamic role in the early reform process. After 1904 the Technologic Branch of the United States Geological Survey, under the direction of Joseph A. Holmes, formerly of the North Carolina Survey, had supervision of the federal government's limited efforts in coal-mining safety. In July 1907 the secretary of the interior, James R. Garfield, had transferred supervision of the coal-mining inspectors in the territories (New Mexico and Indian) to the Geological Survey, suggesting in his order that the survey undertake investigations relating to the nature and extent of mine accidents, particularly those resulting from explosions. The goal, he said, was accident prevention, "without undue hardship to the operators."[21]

At the time of the Monongah and Darr tragedies, the Geological Survey was in the third year of an *ad hoc* study of the causes of mine explosions, financed from a budget item in the appropriation for the investigation of faults which had been stretched to cover the prevention of fuel wastes.[22] Just why the survey became involved in 1906 and who the prime movers were at that time is unclear. Holmes later claimed that as early as 1905 he had recommended that investigations be made into the causes of mine explosions.[23] The director of the Geological Survey, George Otis Smith, indicated his interest in this area to the secretary of the interior and a key member of Congress in the early months of 1906, but at that point, a month before the enormous Courrières explosion in France, Smith hardly viewed coal-mining safety as a high-priority item. His report to the secretary of the interior suggested that investigations into coal gas and coal-dust explosions could be conducted "in connection with this fuel work *without any large additional cost*." The United States, he said, was the only major country not then conducting such tests.[24]

Holmes personally investigated the Monongah situation, arriving in the Fairmont region sometime before December 24, 1907. Although his presence was noted in the local papers, he was not yet well known in the coal-mining safety field and was, in fact, confused with J. W. Holland, director of the museum of the Carnegie Institute.[25] With explosives engineer Clarence Hall of the Technologic Branch (who had arrived soon after the explosion), Holmes did what he could in Fairmont, which was next to nothing since the branch possessed no mine-rescue equipment. The visit of federal experts was, in fact, greeted with some hostility. One coal-mining executive wrote to Fleming from Washington, D.C.: "I saw Mr. Hall, the Government expert, the day he left here, and as you probably know by this time, he is an expert on explosives, but knows little or nothing about mines, or the practical problems of mining."[26] Nonetheless, Holmes's journey to the site of the Monongah disaster was important for coal-mining safety, if only because it emphasized the futility of operating without funds.

One would logically expect that the major force behind the demand for federal appropriations for mine safety would be the miners and their organizations—particularly the UMWA and the American Federation of Labor. This does not, however, appear to have been the case. Although the *United Mine Workers Journal* was aware of the possibility of federal scientific work in the area as early as 1905, this voice of the miners never spoke out for safety appropriations. The president of the United Mine Workers, John Mitchell, made a point of communicating his sympathies for an appropriation to Joseph G. Cannon, Speaker of the House of Representatives,[27] but organized labor in general seems to have waited until the safety work of the Geological Survey had proved beneficial.

The appropriations were obtained from Congress because they had the support of important segments of the mining management community. The most influential group was the one most directly affected by the Monongah disaster: the West Virginia mine operators. With mine operators from Pennsylvania and officials of the Geological Survey, they met in Washington, D.C., on January 8, 1908. The chairman of this gathering, operator William N. Page of West Virginia, asserted the need to determine the causes of explosions and to avoid them. For these ends, the participation of federal and state governments was essential. "A single individual can accomplish nothing," said Page. "We stand to-day in the position of a sick man, and we have the Legislature and Congress as doctors. They want to prescribe medicine when they have not diagnosed the disease." In three official resolutions, those present echoed these sentiments, resolving first that they would support appropriations for research into the causes of mine disasters, second that the causes of disasters were unknown, and third "that the United States Government should take the necessary steps to determine the causes before any attempt is made to apply legislative remedies."[28] The key is in this third resolution: the mine operators present feared ill-advised, hurried, and, they felt, possibly destructive, state legislation. Less than a month before the meeting, Page had written to Fleming on the problem of explosions in coal mines, expressing his interest in correct solutions and his fears of "undigested" legislation.[29] G. H. Caperton, secretary to the group of West Virginia mine operators which later met in Washington, was even more pointed in his views: "It appears that the Governor in his call for a special meeting of the Legislature will incorperate [*sic*] further mining laws in his call. Our committee feels there is to [*sic*] much law now, making the matter of mining a burden and that no further laws could be enacted to help out the situation."[30] Another of Fleming's friends, McKinley, feared "radical legislation" in most coal-mining states and particularly in Ohio, where he anticipated new legislation would be "drastic and burdensome if not almost prohibitive."[31] To say the attitude of McKinley, Caperton, and Page was typical of coal operators of this day would be pure conjecture, but the logic of economics would indicate that this fear of state legislation was widespread. The market for coal was virtually national, with firms from West Virginia to Illinois competing for volume. Costly safety legislation in any one state promised to put operators in that state at a competitive disadvantage. Because of a number of factors—location and accessibility of the coal seams, transportation costs, limited union influence, and low labor costs—the West Virginia operators usually were able to undersell others in the central competitive field; it is, therefore, more difficult to make this case for West Virginia operators than for others. No doubt the operators held a somewhat more subjective view of their competitive situation in the industry, and safety legislation, from their viewpoint, may have been menacing. Operators cautious of such legislation naturally favored federal scientific investigations, the results of which, they felt, would filter down to the states, leaving the competitive situation intact.

One of Holmes's strongest supporters in his campaign for funds was A. B. Fleming. The former governor of West Virginia favored national government research on mine waste and mine-accident prevention and, at Holmes's request, appeared in Washington and testified on the urgency of mine-accident work before the House Committee on Mines and Mining.[32] A publication of Fleming's Fairmont Coal Company recommended that Congress establish a bureau of investigation and information to assist in the study of the conditions under which explosions occur.[33] Perhaps the best indication of Fleming's attitude lies in his answer to McKinley's fears of the upcoming attempts to legislate for safe mines in the states. Fleming said, "I think the operators should be willing [*sic*] to a number of changes which will make the mines safer."[34] Fleming, evidently, had had enough of explosions; he was to be Holmes's resolute ally.

Holmes had begun a strong campaign for safety appropriations for the survey in November 1907, before the Monongah and Darr disasters, and had quickly acquired an unexpected supporter in the normally conservative Pennsylvania Congressman William Dalzell, an influential member of the Appropriations Committee. Dalzell had opposed Holmes in the past but at this time, said Holmes, had volunteered to cooperate "in getting through a *separate bill* placing the Geological Survey on a permanent basis so that no part of its appropriation could be subject to a 'point of order' or be misunderstood or restricted in its application to special areas."[35] In the Senate, Holmes secured support from James A. Hemenway of Indiana and Philander C. Knox of Pennsylvania. It was Knox who replied to those senators who viewed federal mine investigation in the states as an invasion of states' rights with, "Is the protection of human life beyond our jurisdiction?"[36]

Although opposition to the mine-safety section of the appropriations bill was limited, by the time the bill emerged from a House-Senate conference in May 1908, its language was, as Holmes had feared, restrictive. The bill did not specifically authorize safety investigations into causes of accidents other than mine explosions; nor did it allow the survey latitude to engage in mining research except from the perspective of safety. Oscar W. Underwood, senator from Alabama (a state whose coal resources were just beginning to be developed) and supporter of a separate bureau of mines, believed the language of the report too narrow to allow full investigations. James A. Tawney, chairman of the House Appropriations Committee and a member of the conference committee which had emasculated the bill, openly confirmed Underwood's fears. He said the conference committee had intentionally eliminated appropriations for investigations outside the cause of mine explosions, appropriations "that would give the Geological Survey an opportunity to create a permanent organization to go into the business and methods of mining for the purpose of improving efficiency in mining." Tawney based his opposition to broader appropriations on two beliefs: first, that the cause of mine explosions could be quickly ascertained, and thus coal-mine explosions (and presumably coal-mining safety as a whole) did not present a long-range, much less continuous, research problem. Second, Tawney, like others in Congress, feared that permanent mine-safety appropriations for the survey would result in a useless addition to the national bureaucracy. Tawney was also influential in reducing the amount of appropriations under the bill from the survey's estimate of $180,000 per year, and the original bill's provision of $195,000, to $150,000. The next year, when the appropriation came before the Congress for renewal, another attempt by Tawney to reduce funding was foiled by a group of Indiana, Ohio, and Pennsylvania congressmen. In this legislative process, the party issue was raised for one of the few times in all the debates on mining safety, when Congressman John G. McHenry of Benton, Pennsylvania, charged that Republicans alone were fighting the increased appropriations.[37]

Opposition to Geological Survey appropriations for mine safety should not be overestimated. Outside of Congress it was almost nonexistent, and inside it was carried on by a few politicians in influential positions, notably Appropriations chairman Tawney. In general, opposition was extremely limited, partly because of the emotional power of the issue following the December 1907 disasters and partly because of the resolute work of Holmes and the American Mining Congress, an operator organization which supported appropriations through its Committee on the Prevention of Mine Accidents. Securing funds for mining-safety investigations was little trouble for Holmes and the Geological Survey; they had overall support from coal mine operators, metal mine operators (in the American Mining Congress), strong backing from individual operators of authority like A. B. Fleming, no opposition and at times limited support from organized labor, and the strength of public opinion enraged by explosions and backed by the national press. On the other hand, fears of bureaucracy, naive conceptions of the mine-safety problem, and a financial squeeze which arrived on the heels of the 1907 panic were together sufficient to circumscribe the powers and activities of the survey in mine safety; as a result, the survey's potential was limited. Federal involvement, moreover, came slowly and only after incredible numbers of men had died in coal mines. And when federal monies were appropriated by Congress they were hardly adequate to the task. Nonetheless, because of Courrières, Monongah, Darr, and a host of smaller mine explosions and fires, the American public had become concerned and its political system had begun to respond. If federal scientific involvement had not yet been made a permanent solution, in the form of a national department or bureau of mines with statutory authority in the mine-safety field, an important step had been taken by Congress in the temporary funding of the mine-safety work of the Geological Survey.

Except for a brief period following the November 1909 Cherry mine fire, after the spring of 1908 an outraged public played a reduced role in creating the Bureau of Mines. The issue moved into congressional committees and hearing rooms and into the hands of lobbyists for various interest groups. Of particular importance was the alliance of mutual convenience between coal operators and western metal mine operators, functioning through the American Mining Congress. As for the bureaucrats, Holmes remained the consummate politician and the foremost advocate within the federal government of the need for a more extensive mining bureaucracy. Oddly enough, some of the most important opposition to the establishment of a new bureau also came from within the bureaucracy, the product of internal squabbles in the Geological Survey. Otherwise, opposition to the proposal for a bureau or department was limited but strategically placed in Congress, operating on an ideology which stressed the invasion of state police powers by the federal government and the dangers of an enlarged federal bureaucracy. Opponents were never numerically strong in Congress and had no lobbies in Washington. By May 1910, two and one-half years after Monongah and more than four years after Courrières, the various reform forces had coalesced to create the United States Bureau of Mines.

By the twentieth century the bureau as a governmental unit was a time-honored solution to problems requiring national scientific expertise, the roots of the bureau concept antedating the Civil War. The idea began to mature after 1875 with the creation of the United States Entomological Commission in the Department of the Interior and the Bureau of Animal Industry in the Department of Agriculture. Each of these units had one characteristic of the ideal bureau: their center of interest was a problem, not a scientific discipline. The Entomological Commission was established to study the locust, the Bureau of Animal Industry to study cattle diseases. Mine accidents (coal-mining safety in particular and mine safety in general) presented an ideally constructed problem for the problem-solving mechanism of the bureau. As mine-safety work had been developed in the United States Geological Survey up to 1908, however, it lacked other characteristics of the ideal bureau, including the stability to focus on a given line of investigations over a period of years, and the flexibility to shift resources when problems changed. Before 1908 the Technologic Branch carried on mine-safety work only sparingly, using monies appropriated for fuel testing and in the process stretching its statutory power to the limit. After May 1908 mine-safety work had funds of its own, but only for one year at a time, and the wording of the appropriation measure limited the Technologic Branch to investigating explosions. Stability and flexibility thus were not part of the operative framework of the Technologic Branch. From 1908 to 1910, Holmes, acting, in the words of historian A. Hunter Dupree, in the tradition of the "ideal bureau chief," sought to rectify these deficiencies.[38] He fought for an organic act of Congress which would create a bureau of mines governed by a flexible grant of power.

A department or bureau of mines was hardly a new idea in 1908. Efforts to create one began in 1865 and were a regular feature of the political scene. Of the major attempts, one was rebuffed in 1881, another in 1898, and from 1900 to 1907 a number of bills calling for a department or bureau of mines were presented in the House and the Senate. With few exceptions, these bills did not mention mine safety, much less coal-mining safety; their purpose was to promote and encourage mining and metallurgy, to aid in the development of the nation's mineral resources. They failed to become law largely because they were supported only by limited groups outside of Congress, particularly western mining interests. Bureaucratic pretensions, however, also played a part. Charles D. Walcott, for many years director of the Geological Survey, was particularly jealous of his agency's jurisdiction. In support of one measure which would have created a division of mines and mining within the Geological Survey, Walcott charged advocates of a separate department with having "lost sight of the fact that an existing organization, the Geological Survey, includes among its duties many that could be done by such a department."[39]

Support for a bureau of mines became widespread in late 1907 and early 1908, both without and within the Sixtieth Congress. Some one thousand newspapers around the country called for some form of federal action to prevent mine accidents.[40] These newspapers reflected the strong support for the concept of a bureau among every major group with interest in the subject—miners, operators, conservationists, scientists. In congressional hearings, operators were virtually unanimous in support of a bureau of mines. G. W. Traer of the Illinois Coal Operators' Association spoke for a committee of coal operators which was unanimously in favor of a bureau. W. W. Keefer, representing the Pittsburgh operators at the hearings, emphasized the coal-dust problem, noting: "We are free to confess that we are unable to agree with other people as to any of the causes and conditions that have brought about these horrible explosions; we would like to be advised; I believe everybody has the same general view of the problem." A well-known coal-mine operator, S. A. Taylor of Pennsylvania, saw the handling of dust and explosions as basic to the bureau issue. "These are the things," he said, "that this bureau ought to take up and analyze and disseminate the results broadcast among the mining industry."[41]

As in the struggle for appropriations for the Technologic Branch, Holmes's strongest ally in his quest for a bureau of mines was operator A. B. Fleming, who worked throughout 1908 and 1909 to secure passage of the bureau bill. Fleming corresponded regularly with Senator Nathan Scott of West Virginia in a successful effort to secure his support for the bureau. Holmes provided Fleming with information to use in writing Scott and other members of Congress and invited the former governor to the 1908 House hearings, where he proved to be one of Holmes's most important witnesses. Before the Committee on Mines and Mining, Fleming acknowledged that neither the largest operators nor the states could afford to carry on the kind of scientific investigations requisite to safe mining.[42]

The operators held back only when their interests were directly, and they thought adversely, affected by a bureau bill. Such was the case with Congressman John G. McHenry's proposal, which not only empowered its projected bureau to investigate the causes and effects of all accidents in coal mines but also provided for a tax of one cent per ton on all coal mined in the United States, to be collected from owners and operators and to be used to alleviate the suffering of the families of those killed or injured in mine accidents. It empowered the bureau to examine the books of any coal company to ascertain compliance with this provision. Although the McHenry bill enjoyed considerable popularity among miners, their representative in Congress, William B. Wilson of Pennsylvania, opposed the bill because he felt it might jeopardize the campaign for a bureau by encouraging operator resistance.[43] There was considerable logic in Wilson's position, for most coal operators balked at the hint of federal interference with mining operations in the McHenry bill. "The work of the bureau," said operator Traer, "should not have anything to do with the extension and supervision of practical mining operations."[44]

The most influential organization of operators and their only national organization lobbied hard for a bureau of mines but cared little for coal-mining safety and not much more for safety in the metal mines. Originally called the International Mining Congress, the American Mining Congress (AMC) held its first annual meeting in Denver in 1897, drawing its constituency from operators, owners, prospectors, and miners from states west of the Missouri River and from Mexico. Although miners were allowed membership under the constitution, in practice the organization was dominated by mine operators and its policies reflected this dominance. Its charter paid lip service to nonmetallic mineral resources (i.e., coal), but in reality the organization was a sectional one dedicated in general to the development of western metal mining and metallurgy and their allied industries and specifically to the creation of a department of mines and mining. For the first ten years of the organization's existence, this particular goal was singularly unrelated to safety. Spokesmen for the AMC emphasized how a bureau might aid in the production and discovery of rare minerals, such as uranium, tungsten, and gold.[45] J. H. Richards, long-time president of the organization, captured the essence of this view in his address to the 1904 convention:

> We therefore affirm that if a Department of Mines and Mining could broaden the markets for the products of our mines by intelligent investigation and official action; if it could diffuse among prospectors and miners in practical form the scientific information which would be so useful to them; if it could afford them cheap and perfectly reliable facilities for classifying and assaying the infinite variety of ores found in our extensive mineral districts; if through a revised, simplified and uniform system of mining laws and a judicious control of mining corporations in the interests of the working miner, the investor and the general public, it could lessen that element of friction and speculation which to-day in the opinion of so many condemns mining both as an occupation and an investment— . . . then it must be apparent to all that such accomplishment would create a new atmosphere and a new hope, not only throughout the mining world, but that the salutary effects would react in the commercial and industrial world, and that every railroad office, every bank, every factory and every farm would feel the stimulus and reap the benefit.[46]

This was the language of business, of investment, of progress and development in the mining industry. Not until 1906 would the first paper on mine safety be delivered at the annual meeting.[47]

By 1909 the American Mining Congress had been transformed in several important ways. It had become, through sheer necessity, an advocate of safer mines, with its own committee, staffed with some influential coal men, to deal with prevention of mine accidents, and its lobbying had begun to reflect this conversion. AMC leaders approaching the interior secretaries, Garfield and Richard Ballinger, now at least mentioned mine safety. As of 1910 the AMC nominally and superficially accepted the safety issue and was using it for its own purposes. In a related transformation, the AMC had tried to broaden its appeal and bring eastern coal operators into the organization. Recognition of this need came in 1905, and by 1908 the annual meeting took place, symbolically, in Pittsburgh. Finally, the congress had changed its goals. Old demands for a department of mines and mining, usually emphasizing the desire for equality with agriculture, were gone; in their place was the demand for a simple bureau of mines. It was not easy for an organization which had previously feared that its industry would be "pigeon-holed in a bureau" to forsake its departmental dreams. In 1907, however, Richards officially pronounced an end to those dreams when he asked the annual meeting to be satisfied with a bureau.[48]

The AMC exercised its influence through a strong lobbying effort headed by the congress secretary, James Callbreath. Callbreath handled political correspondence, testified at the 1908 House hearings, and worked at getting such important mining men as A. B. Fleming interested in the work of the AMC. When the Committee on Federal Legislation, appointed following the 1909 annual meeting, was debilitated by the death of its chairman, it was Callbreath who went to Washington to fill the void.[49] He was also informally in charge of a committee of three set up in 1904 to guide through Congress a bill creating a department of mines and mining.[50] This committee maintained communication with key senators and congressmen and national party leaders and called on the major parties to write the demand for a department of mines into their national platforms for the 1908 elections. Representatives of the AMC at one time or another talked or corresponded with Presidents Roosevelt and Taft and their interior secretaries, Garfield and Ballinger; all four were in favor of establishing a bureau of mines.[51] Congress president Richards reported a discussion with Roosevelt sometime in late 1906 or early 1907: "He is a very strong character, and is like a steel trap. He said: 'What do you mining men want?' I said: 'We want results; we don't care what you name it.' He said: 'I will recommend a Bureau of Mines in my next message. I have a part of it outlined already. Will that suit you?' I said: 'That is all we ask. Good-bye, Mr. Roosevelt.' Quick work by a quick man, and that is the way we expect that bill to pass, or else they will hear from us in Washington again."[52] If Richards was too optimistic, it is nonetheless clear that the American Mining Congress could command the attention of the highest public officials.

UMWA support for a bureau of mines, unlike that of the American Mining Congress, was based almost entirely on the desire for safer mines. The extent of labor's interest in a bureau is, however, open to question. William B. Wilson delivered the following speech on the floor of the House in April 1908:

> Tell me in martial measures of the Charge of the Light Brigade; point with patriotic pride to the farmers who fought with the soldiers at Concord and Bunker Hill; repeat to me the story of the mighty conflict between the Blue and the Gray, when the "flower of American youth yielded up their last full measure of devotion" in defense of their respective flags, and my blood will thrill with patriotic enthusiasm and tingle through my veins in sympathetic response. Yet, with all my admiration for the heroes of the battlefield and their wonderful achievements in support of the rights of man, it does not equal my love, it can not measure my devotion to those sturdy sons of toil who, uninfluenced by the enthusiasm of numbers, without hope of present reward or future glory, deliberately enter the dark and dangerous caverns of the mines to carry relief to their suffering fellow-men or perish in the attempt. It is for the benefit of men of this character that I appeal to you to establish and equip a bureau of mines and mining. (applause)[53]

As public relations, Wilson's address is explicable; as proof of labor's emotional attachment to mine safety as it relates to a bureau of mines, it is inadequate, for there is little in the labor sources to confirm this enthusiasm. Far from exercising a leadership role in the national politics of mine safety, the mine-workers union was a cautious follower. The organization did little more than go through the motions. The *United Mine Workers Journal* had shown interest in a federal department as early as 1902, but it soon dropped the subject (which was not, at the time, connected with safety) and did not resurrect it for about five years. The national organization of the UMWA did not evince interest in a federal department or bureau until 1907, when action began from the top, with the union's National Executive Board recommending establishment of a bureau at its June 25 meeting. Even here, the union assumed a passive role, for the board took action only after Holmes had appeared before it that same day and explained the major goals of the mine-safety investigations of the Geological Survey. Six months later the National Convention confirmed the board's action by a unanimous rising vote.[54]

The UMWA's political apathy emerged from a recognition of the strength of business in the national politics of mine safety and from a general awareness that any institution emerging from these politics would fall well short of the union's standard. The radical McHenry measure was the clear favorite of the miners, the choice of the national convention, John Mitchell, and John Walker, president of the Illinois UMWA and a spokesman for the miners at the 1908 hearings.[55] But even William Wilson refrained from supporting the measure in Congress, and with the McHenry bureau out of reach, the mine workers apparently lost some of their enthusiasm for the project, content to work at keeping the proposed bureau in the Interior Department and out of the Department of Commerce and Labor.[56] Labor's political diffidence in this area might also be the product of reasonable suspicions about placing safety regulation in the hands of mining engineers whose national organization, the American Institute of Mining Engineers (AIME) was traditionally hostile to organized labor and social reform. The UMWA apparently had no Washington lobby until March 1910, when its president, Thomas Lewis, appointed a committee of three to look after national legislative matters. On May 14, 1910, two members of that committee received a hearing with President Taft along with other miners, operators, and politicians. The bureau bill, already passed by Congress, was signed by the president two days later.[57]

The state of the nation's natural resources was a common concern of the age, and politicians, mine operators, labor leaders, and others envisioned the Bureau of Mines in part as a conservation agency. On the floor of the House, the proposed bureau was defended for the impact it would have on use and allocation of resources. Congressman Burton French of Idaho spoke of the bureau's potential for eliminating coal waste caused by burning and poor production techniques, and a fellow westerner, Congressman William Englebright, emphasized the need to protect the coal supply of the United States for future use.[58] During the hearings, a Pittsburgh operator boldly suggested that the problem of explosions would be incidental to the bureau's operations: "It will have many other functions, one of which, I understand, will be to look after the waste of the mineral resources of the United States, not only of coal, but of all the other mineral resources. We have only recently awakened to the fact that the country has almost impoverished itself as to its timber supply, and now every force and influence of the Government is being brought into play to see what can be done."[59] Occasionally, the natural resource aspect of conservation was directly related to the safety issue: unsafe mining techniques were attacked as wasteful of natural resources as well as lives. The two varieties of conservation—human and natural resource—were linked especially closely in the case of the mine fire, a deadly and costly phenomenon. Here the Cherry mine fire may have been a direct influence. The American Association for the Advancement of Science, one of several scientific groups to participate in the campaign for a bureau, separated the conservation themes but employed them both. Holmes, himself a conservationist, was also an intimate friend of Gifford Pinchot, a man of considerable influence within conservation circles. Holmes often used Pinchot's contacts to obtain endorsements for the bureau from conservation organizations like the National Irrigation Congress and the Trans-Mississippi Commercial Congress.[60]

In addition to the pleas for conservation of human life and natural resources, a bureau of mines was urged upon the Congress because it would encourage uniformity of state legislation.[61] Attempting to ease the fears of constitutional conservatives, Congressman George A. Bartlett of Tonopah, Nevada, spoke to this point: "There is not a single provision upon which to hang a fear of Federal usurpation of the rights of the States. . . . the purpose of this act is to aid and assist in bringing about, by investigation and suggestion, a uniformity of State legislation that will lessen the terrible loss of life incident to mining, and to further aid in the development of the vast mineral resources of the country."[62] Operators and union chiefs also recognized uniformity (and its counterpart, standardization) as a valid objective, and the organization established precisely to pursue uniformity of state legislation, the Mine Inspectors' Institute of America, came out in support of a bureau of mines. The political influence of the inspectors, however, was limited by the small size of their organization, which prevented the institute from employing a federal lobby.[63]

Unlike the bureau's proponents, the opponents had no organized public support and no lobbies. Nonetheless, each time bureau bills received serious consideration, in 1908, 1909, and even 1910, there was some kind of opposition. Perhaps the greatest obstacle to the creation of a bureau of mines in the early months of 1908 was lack of support from the survey's director, George Otis Smith. Two contemporaries made this charge. Gifford Pinchot wrote in his diary on December 9, 1907, two days after the Monongah disaster: "G. O. Smith wanted to talk about Bu Mines wh. he opposes."[64] During hearings early the next year, Congressman Albert Douglas of Ohio implied that Smith had little interest in a bureau of mines and instead wanted the Technologic Branch of the survey to continue to handle mine-safety investigations. "The Geological Survey," he said, "is going after this work and trying to get it to develop their own importance."[65] Smith left some indirect evidence of his opposition in the form of three memoranda dealing with new possibilities for mine-safety work. A bureau of mines was not even mentioned, much less advocated. In one memorandum, Smith specifically suggested that the investigations could be conducted in connection with the fuel work of the Technologic Branch.[66] When, in 1907, Holmes began his campaign for a bureau by attending mining conferences and other meetings in an attempt to build a base of support for the legislation, Smith wrote Holmes requesting that he "refrain absolutely from taking any further part whatsoever in recommending through resolution or otherwise, the establishment of a Department of Mines or Bureau of Mines except as you may be authorized or instructed by me."[67] Whatever the basis for his attitude, be it personal or political, Smith did not want Congress to create a bureau of mines, and his reluctance to work for the bill may have been sufficient to route the first session of the Sixtieth Congress in the direction of additional appropriations for the Geological Survey.[68]

Smith had some support in Congress, however unintentional, from a small but determined coterie of conservatives, led by James Tawney, which viewed a new bureau of mines as a symbol of that deadly enemy bureaucracy. Tawney feared that the United States was "unconsciously drifting toward a highly organized, bureaucratic form of Federal Government."[69] James Slayden of San Antonio echoed Smith's views: "I object to the creation of any more bureaus of the Government. If this work is important enough to have it done, then, sir, it might well be delegated to the Geological Survey, an important bureau in the Department of the Interior. . . . it means a multiplying of officials; it means tremendous increase of expense. . . . This proposed bureau will duplicate work being done by the Geological Survey . . . particularly there should not be a multiplication of high-priced officials whose services are not required by the public service or in the interest of humanity."[70] In the Senate, this viewpoint was defended by Alexander Clay of Georgia, Moses Clapp of Minnesota, and Jacob Gallinger of New Hampshire. Idaho Senator Weldon B. Heyburn agreed the bill would accomplish "only an expensive addenda to the administration of the law," but he would not oppose it. "It is," he said, "one of those fads that take positions of public importance at intervals and the only way to meet them is the manner in which Tom Sawyer appeased the cat's curiosity for pain killer."[71]

Concern for the traditional role of the states in mine inspection and disaster investigation was common to bureau proponents and opponents. States' rights was here not a reactionary doctrine but a consensus view. Senator Nathan B. Scott of Wheeling, West Virginia, one of the hardest workers for a bureau of mines, said: "I do not want to see the General Government do more than it properly can do, than the States themselves. The foundation of our Government is local self-government."[72] Representing the Technologic Branch of the survey at the 1908 meeting of the Mine Inspectors' Institute, engineer Clarence Hall emphasized that the federal government did not intend to make inspections or regulations or otherwise interfere with the rights of the states. Holmes and Smith agreed, as did congressional advocates of the bureau who assured their colleagues that the bill did not involve an infringement upon the police powers of the states and that any attempt to invest the federal government with power to control mines within the states would be unconstitutional and unjustifiable. During the House hearings, John Walker of the mine workers had suggested federal legislation and enforcement as solutions to some of the mine-safety problems, but his ideas were atypical. Advocates of federal inspection and investigation were few. The preponderant view was that expressed in all the major bills, even those which failed to become law. Certainly H.R. 20883, the major subject of consideration in both the first and second sessions of the Sixtieth Congress and the bill passed by the House in 1908 and 1909, contained nothing to indicate that the bureau would have powers of inspection or of investigation of mine accidents.[73]

A Senate report favorable to H.R. 20883 nonetheless stated that "many advocates for a department or bureau of mines mention the need for Federal inspection of mines."[74] Unnecessarily alarmed, Senator Henry Teller of Colorado objected to consideration of the bill and it was held until the next session, which began in late 1908. Teller later claimed that the bill would mean "an absolute interference with the mining operations of this country, particularly with the mining of precious metals . . . we do not need any man to come from Washington out there to tell us how to develop our mines, or, in the words of the bill, how to 'promote and develop the mining industries of the United States, to make a diligent investigation of the method of mining, the safety of miners,' and so forth."[75] Like Tawney in 1908, Teller objected to a bill which went beyond the immediate goal of coal-mining safety. And, like Tawney in 1908, Teller in 1909 was forced to use blatantly deceptive arguments to bolster his opposition to the bill. Ignoring the American Mining Congress, Tawney had argued that there was no demand for this legislation from the precious-metal mining states. Teller also considerably overstated the case when he said: "I deny that there is any demand among the miners of this country for this kind of legislation." Both men were in small minorities; both wielded great power in their committees. In their separate ways they succeeded in delaying for two years the passage of a bill which was opposed by so few persons or groups that Senator Francis G. Newlands of Nevada could say: "I have never yet, to my knowledge, received any communication in opposition to the organization of a bureau of mines."[76]

When the second session of the Sixtieth Congress opened, H.R. 20883 remained in Senator Charles Dick's Committee on Mines and Mining, and as Holmes put it, "senatorial 'courtesy' and senatorial inertia [allowed] one determined old man to prevent important legislation." "In this case," Holmes continued, "Senator Teller (of Colorado) seems to be the barrier. Whether we can get the bill passed in spite of his opposition I am unable to say."[77] Lobby pressures were sufficient to get the bill out of committee, but it reached the floor of the Senate very late in the session and was "talked to death in the final hours of the Senate."[78] "The American House of Lords," said the *United Mine Workers Journal*, "is fast becoming an exact counterpart of the English House of Lords."[79]

Important new conditions greeted the introduction by congressman and coal operator George Huff of Greensburg, Pennsylvania, of H.R. 13915, to establish a bureau of mines in the Department of the Interior, on December 10, 1909. First, Teller had retired from the Senate. Second, a new climate for the bureau had been created by the Cherry, Illinois, mine fire of November 13, 1909, and the publication, in early 1910, of Crystal Eastman's book *Work-Accidents and the Law*. The latter, while not intended to be an exposé of coal-mining safety conditions in the bituminous coalfields of western Pennsylvania, was precisely that.[80] The intensity of feeling invoked by the Cherry disaster is revealed in an exchange between socialists Adolph Germer and Eugene Debs. Germer, who was in Cherry soon after the fire, wrote Debs that "every indication pointed to the flagrant neglect of the lives of these men. With the least equipment for dealing with perils of this kind, the lives of these miners could have been saved." Germer condemned the coroner, the state's attorney, and the company. Over 300 lives, he said, "have been snuffed out through capitalist greed." Debs replied: "Note what you say about the Cherry disaster. It was undoubtedly murder in the first degree. My blood boils when I think of it and my heart bleeds for the widows and orphans."[81] Senator Dick brought the matter home to the Senate in a denunciation of the restrictive provisions of existing national mine-safety legislation:

> Nothing better illustrates the inadequacy of the investigations already authorized by Congress through the Geological Survey than this

experience at the Cherry Mine. The experts of the Geological Survey went to this mine on delayed telegraphic notice, on the supposition that it was probably a mine explosion. As a matter of fact, it was not an explosion, but a mine fire, and under the wording of the appropriation, which limits the investigations to "mine explosions," they had no right to go, and after they arrived at the Cherry Mine, under a similar strict interpretation of the act, they had no right to remain or to aid in the rescue work, nor have the experts of the Government, under existing law, any right to aid the mining industry in any of the ways mentioned above looking to the prevention of mine fires, better systems of mine signals, and better methods of mine rescue work.[82]

Coupled with the impact of the Cherry fire and the influence of Eastman's work was a willingness on the part of the Sixty-first Congress to amend the bill to suit its critics. Dick's Senate Committee on Mines and Mining added a section explicitly disavowing any intent to grant the bureau powers of inspection or supervision.[83] In doing so, however, the committee inadvertently eliminated federal inspection not only in the states but also in the territories, where federal personnel had been inspecting mines since 1891. When this defect was corrected, the bill was complete. Thomas H. Carter of Montana, who had previously opposed the measure on constitutional grounds, now argued in favor of the bill as a way to achieve uniformity in mining-safety legislation. Another former critic, Jacob Gallinger of Concord, New Hampshire, also spoke for the bill. With opposition virtually eliminated, the Huff bill passed the Senate on May 2, 1910, and was signed into law by President Taft on May 16, 1910. "Colonel Huff," said the *United Mine Workers Journal*, "regards the passage of this bill as the crowning effort of his congressional career."[84]

The permanent directorship of the new Bureau of Mines lay vacant from July 1, when the act went into effect, until early September, when President Taft selected Holmes to fill the post. Behind the unusually long delay was politics, altogether not a promising beginning for a new bureau but revealing of the nature of the decision-making process and of the intense interest which even a minor appointment could invoke. It would be all too easy to interpret Holmes's appointment as a victory for various establishment elements—the coal-mine operators, in particular. The appointment was that, to be sure; but that may not be the most important lesson to be learned from it. In fact, the history of the appointment indicates the high degree of influence exercised by the existing government bureaucracy and the extent to which corporate interests could not manipulate the political process.

Holmes was apparently in an ideal position to become the first director of the bureau. It had, after all, been established to save lives, conserve natural resources, and contribute to the development of the nation's mineral resources, and in all three areas Holmes had considerable experience and had demonstrated dedication. Born in 1859, he was graduated from Cornell with a Bachelor of Science degree in agriculture in 1881. After graduation, Holmes was appointed professor of geology and natural history at the University of North Carolina, where he actively and successfully campaigned for the establishment of a state geological survey. When the legislature created the North Carolina Geological Survey in 1891, Holmes became its director as state geologist and continued to reside at the University of North Carolina for a dozen more years. During this period Holmes agitated for good roads in North Carolina and acquired a national reputation as a roads expert. More important for his future career, Holmes conducted a broad examination of North Carolina's mineral resources and was able to bring domestic and foreign capital into the state for development of mines, forests, fisheries, and water-power resources.[85]

In 1903 Holmes became chief of the Department of Mines and Metallurgy of the Louisiana Purchase Exposition at Saint Louis, and from this time on his strongest ties were federal ones. At the exposition Holmes concentrated on demonstrating fuel economies and was so successful that Congress in 1905 authorized an investigation into fuels and building materials under the United States Geological Survey at Saint Louis and Washington, D.C. Responsible for this investigation as geologist in charge of fuel investigation for the survey, Holmes and his department investigated the relative values of coal, lignites, oil, and other fuels, work that led to the employment of more economic and efficient methods of purchase and use. Because of these efforts, Holmes attained a national reputation for conservation, later validated by his activities in promotion of coal-leasing legislation and by his selection in 1908 as one of the four secretaries of the National Conservation Commission, a governmental agency.[86] Holmes became expert in charge of the Technologic Branch of the Geological Survey when it was created by the secretary of the interior in 1907 (absorbing the work of the fuel division) and continued in that capacity until his appointment as bureau director. The step from the Technologic Branch to the bureau was a natural one, for the branch was to constitute the basis of the bureau. Personnel, equipment, and investigations of the Technologic Branch would all transfer to the new agency.

Holmes also had the support of key interest groups, with particular strength among eastern coal operators. Four days after the bill creating the bureau was signed, a group of coal operators met in Washington and united in a request to the president that Holmes be placed in charge of the bureau. In mid-summer, A. B. Fleming led another delegation of Illinois, Ohio, and Pennsylvania operators to Washington to press Holmes's candidacy. One West Virginia operator, appreciative of Holmes's services to the industry following the Monongah disaster, wrote: "We, who through our great disaster in 1907, came in contact with Doctor J. A. Holmes, . . . know that he is, by disposition, theoretical knowledge and experience, thoroughly qualified for this position."[87] Holmes also had the endorsements of the American Association of State Geologists and the American Association for the Advancement of Science. Although the American Mining Congress did not endorse Holmes, this appears to have been a matter of policy rather than a sign of disfavor. Holmes's work at the bureau was later loudly applauded by the western mining industry.[88] Taft, too, was ultimately to acknowledge all this. Having made the appointment, he wrote to his secretary of the interior: "The truth is that in searching for a competent man I do not find any one whom I am willing to appoint and take the responsibility of turning down a man who comes to me recommended so highly by both the labor organizations and the mine owners and superintendents."[89] Possessed of great energy and enthusiasm for his work, Holmes corresponded regularly with governmental officials, friends, and acquaintances and attended conferences and conventions of many organizations of which he was a member and many of which he was not. He often used his private funds to supplement the state's appropriations for geological investigations while with the survey in North Carolina, and it was there that he voluntarily reduced his own salary, already low, so that he could increase that of his assistant. Gifford Pinchot described Holmes as "one of the best men I ever ran across."[90]

With such strong qualifications and support, why was the appointment ever in doubt? For one thing, Holmes's candidacy was caught up in a backlash against the cult of the expert and progressive ideas of efficiency which captured both miners and inspectors and filled the columns of the *United Mine Workers Journal*. This backlash was premised in part on a lack of appreciation for the conclusions of the new science of mine safety and in part on a reasonable desire to avoid what until then had been common practice—the dictation of mining procedures by those who had little knowledge of what went on inside a coal mine. Although it never explicitly came out for or against Holmes, the *Journal* clearly favored a "practical" coal-mining man, a miner or perhaps a mining engineer, for head of the new bureau. The explosions in 1907 and 1908 brought *Journal* hostility to a peak. "Long-haired theorists" in charge of coal-mine investigations were denounced; the need, said the *Journal*, was for "some good, level-headed practical coal miners in charge." Both experimental work and rescue work of the Technologic Branch were criticized; not enough money was being spent "in a practical effort to get at the cause of all our suffering." The "long-haired college professor," with his "multitude of theories and a molehill of practical knowledge," also was subject to attack.[91] Joining in the criticism was Andrew Roy, at age seventy-four a candidate for the bureau directorship in 1908, who said he "would regret to see any of the so called experts appointed. . . . They are good at writing well rounded periods, but so far as subtarranean [*sic*] knowledge and experience goes they are sadly lacking."[92] Although some of the scientific ignorance upon which these attitudes were based had vanished by 1910, the *Journal* held firm and asked for a practical, competent coal miner as head of the bureau so that "it may not be made a farce in the hands of some professional expert."[93] A number of the state delegations of the Mine Inspectors' Institute—particularly Illinois, Ohio, and Indiana—were dominated by practical mining men and so had similar views. Of the mining journals, only *Mines and Minerals* had some sympathy with the demands for a practical miner; and that journal found Holmes satisfactory, acknowledging that to select either a coal- or metal-mining engineer would alienate the other group.[94]

The anti-expert attitude not only detracted from Holmes's candidacy but it also furthered that of David Ross of Illinois. According to the *Journal*, Ross, then secretary of the Illinois Bureau of Labor Statistics, a former UMWA organizer, and an active member of the Illinois Miners' Protective Association, "still has a very warm side for the boys with whom he once toiled 'down in a coal mine underneath the ground.' "[95] The Ross candidacy still might never have developed into a serious challenge to Holmes had not the nation's mine inspectors split on the director issue. Officially eschewing participation in the appointment controversy because the "candidates are being urged for endorsement by their respective friends" and because the Mine Inspectors' Institute of America did not want to "participate in acrimonious debate which may defeat the good purposes and efficiency of the organization," in reality the inspectors chose not to participate because they could not agree. While most of the inspectors favored Holmes, a significant minority from Illinois felt Ross was better suited to their purposes and views. An attempt by the Illinois delegation to put through the convention a resolution in favor of Ross was defeated by a combination Pennsylvania and West Virginia delegation led by John Laing, chief mine inspector of West Virginia.[96]

Inspector disaffection with Holmes had its origins in a less than perfect relationship between the Illinois inspectors and the Technologic Branch of the Geological Survey, the source of conflict evidently lying in either disappointment with the rescue work of the survey or in jealousy between state and federal investigators.[97] Possibly some inspectors (and some operators) feared that the bureau under Holmes would undertake functions they felt properly belonged to the states. Holmes, at least, did his best to discourage such talk, emphasizing in a letter to John Mitchell that there would be no friction with state authorities if he were director of the bureau. "The only part of the act establishing the Bureau of Mines for which I am responsible," wrote Holmes, "is the section which provides that the employees of this Bureau shall have nothing to do with the inspection of mines in any state." To the states and private corporations, he emphasized, should belong the "burden of inspection and other local problems."[98] Holmes was also unpopular with the more radical mine inspectors for his opposition to popular election of inspectors. Here at least he had the support of Laing, the product of the political influence of the West Virginia operators. David Ross supported the elective concept and for Laing and others, his candidacy evoked fears of labor control of mine inspection. The mine inspectors of Ohio, Indiana, Illinois, and the majority of western states, wrote Laing, "are labor agitators and their sole desire seemed to be to so regulate the work of the Mine Inspectors Institute . . . that it would be controled [*sic*] absolutely by politics under the direct supervision of the labor organizations."[99] Perhaps for this reason Laing spent a week in Washington pressing the Holmes candidacy. The Fleming-Laing coalition must have had some impact, for by the middle of August, Ross was no longer being seriously considered for the directorship.

The most critical opposition to Holmes came from within the Taft bureaucracy. Prompted by George Otis Smith, head of the Geological Survey, the administration was worried about Holmes's past (and thus future) loyalty to his superiors in the government. The origins of this problem can be traced to 1907 when Holmes, as expert in charge of the Technologic Branch of the survey, was beginning his campaign for a bureau of mines. Smith objected in writing to the principle of Holmes influencing legislation in which he (Holmes, although really Smith) was directly concerned; and Holmes, after claiming that he was unaccustomed to occupying a subordinate position, continued to lobby assiduously for a bureau of mines. In 1910 Smith found that Holmes was serving as a member of the Committee on Resolutions of the American Mining Congress. The Geological Survey director, who had prohibited that particular kind of service, then concluded "that at best the Technologic Branch has tendered the Director of the Survey only the most perfunctory allegiance. Your methods both of administration and of influencing public opinion are directly opposed to the policy now in force, . . . and any further direct violations of my orders or indirect regard of my expressed wishes will be made the subject of immediate reference to the Secretary of the Interior with recommendation."[100] James Garfield, former interior secretary and a great admirer of Holmes, felt Smith "had not treated Holmes well. He has been jealous of Holmes' influence and work."[101] When *Coal and Coke Operator* learned that Smith, as the bureau's acting director, had assigned Holmes to subordinate duties, it accused Smith of "professional jealousy" and called him a "mediocre official."[102] Motivated by professional jealousy or not, Smith apparently felt that the new bureau should in some way be subordinated to the survey. "The Geological Survey," noted the *Engineering and Mining Journal*, "evidently desires the Bureau of Mines to be a tail to its kite."[103] Smith, of course, saw Holmes as the culprit and resented both his lobbying efforts and his conception of the proper role of the Technologic Branch and, indeed, his whole concept of administration. Holmes wanted to push the branch rapidly in several very different directions. Smith felt the branch, and the new Bureau of Mines, should concentrate on mine safety and postpone metallurgical and mineral investigations until appropriations were larger. "The danger which threatens the new Bureau," said Smith in July 1910, "is the same which has hampered the Technologic Branch of the Geological Survey, namely, the tendency to take up new lines of investigations before results are fully attained along other lines earlier undertaken." Smith also charged Holmes with failure to keep sufficient control of those working under him.[104]

For Smith this was serious business. He had tried, and failed, to prevent Congress from establishing a bureau of mines. Now, to protect the prerogatives of the survey and his personal influence and status, he was prepared not only to oppose Holmes but to present his own candidate. His choice fell upon Edward W. Parker, since 1891 a loyal survey employee and in 1910 the survey's chief statistician. In recent years, Parker and Smith had worked closely together in instituting survey reforms in accounting and business methods. Although Parker had done some scholarly research and writing on coal mining, his knowledge of mine safety and his ability to run a bureau were suspect quantities. This deterred Smith not at all. While claiming that the impetus for Parker's candidacy came from other officials at the survey, Smith inaugurated a mail campaign in Parker's behalf, requesting support from influential mining men across the country.[105]

This ambitious bureaucratic venture into appointment politics was to fail, not because Parker was unattractive to major interest groups—Laing and Fleming and presumably other operators would have tolerated his appointment—but because events beyond Smith's control would soon make the chief statistician unacceptable to the administration. On July 2 Parker was accused by A. J. Chipman, a retired survey receiving clerk, of having expropriated a small quantity of public funds—specifically, of having attended his father's funeral at public expense. Under normal circumstances, the brief investigation which followed would have been sufficient to clear Parker. But with the Ballinger-Pinchot affair already in the public eye, the administration could ill afford to press a candidate even slightly tainted by scandal. Guilty or innocent, Parker was a political liability in 1910. Parker apparently failed to persuade Taft to the contrary in a disastrous mid-July interview with the president. Faced with the Chipman accusation, Smith reconciled himself to working with Holmes.[106]

Sometime between July 15 and August 15, the administration found itself with no respectable candidate except Holmes. Still he was not appointed. Again, Holmes's reputation for insubordination came back to haunt him. At issue this time was Holmes's relationship to Richard Ballinger, interior secretary. Holmes was a Democrat, a Roosevelt appointee, and a close friend of Gifford Pinchot, James R. Garfield, and F. H. Newell, all of whom testified against Ballinger during the Ballinger-Pinchot hearings in the early months of 1910. The Washington press speculated that Ballinger was opposing Holmes's appointment because Holmes had played a part in the campaign against him.[107] Whatever his personal and professional inclinations, Holmes had not contributed to the anti-Ballinger campaign, perhaps knowing that any involvement here would crush his chances of appointment to the directorship. But the rumors themselves did damage, no doubt contributing to Ballinger's and Taft's impression of Holmes as a difficult man to control.

Holmes did his best, before and after his appointment, to assure the administration of his loyalty. In late July, when articles appeared attributing the appointment delay to Holmes's disloyalty to Ballinger and Taft, he spoke several times with Donald Carr, Ballinger's secretary in Washington, D.C., attempting to convince Carr, and through him Ballinger, that he neither was responsible for the articles nor liked their content. When, on the day of the appointment, the *New York Herald* again claimed that Holmes was antagonistic to Ballinger, Holmes saw Carr and expressed regrets. Holmes, wrote Carr, "has reiterated several times the statement that he is not one of our enemies."[108] Three days later Holmes wrote to Ballinger, assuring the secretary that "in the discharge of the duties assigned me as Director of the Bureau of Mines, I shall have no higher aim than to properly carry out the purpose of Congress in establishing the Bureau, as that purpose shall be interpreted by the Secretary of the Interior, and to cooperate loyally with my associates in this and other bureaus in endeavoring to carry out the policies and plans of the Department and of the Administration."[109] Through John Mitchell, who had the ear of the president, Holmes attempted to allay these and other suspicions about his loyalty. Ballinger had misunderstood his purposes and actions; he would have no friction with the secretary. "The reports you referred to concerning my excessive activity in legislative matters and tendency to insubordination," Holmes wrote, "are alike based on an exaggerated misinterpretation of the facts and are without any basis."[110]

On this key point of loyalty the administration was ultimately reassured. Taft's interview with Holmes in August dealt almost entirely with the subject of Holmes's loyalty to Ballinger. Taft was prompt in assuring Ballinger that "he promises the utmost loyalty to you." Referring, evidently, to Holmes's legislative activities under Smith, Taft added: "I think we can make his activity an asset of usefulness in the administration."[111] Ballinger wrote Taft following the appointment:

> My greatest concern has been that I feared Mr. Holmes was thoroughly inoculated with the idea of bureaucratic ascendency, and he bears the reputation of intermeddling in legislation beyond the control of his superiors.
>
> I am pleased to note in your letter that he has agreed to subject him-

> self to authority and to work with loyalty to the head of the Department. I shall endeavor on my part to make it easy for him to keep his promises.[112]

Ballinger did not oppose Holmes, and the final obstacle to his appointment was removed.[113]

If, in view of his difficulties with Smith and his association with the opponents of Ballinger, Holmes was himself a liability, the alternatives —Ross and Parker—had more serious deficiencies. Ross had support only from limited groups of miners and mine inspectors; Parker, whose base of support was narrower but similar in kind to that of Holmes, had no experience as a bureau chief and alleged fraud marred his background and left the administration, already suffering a crisis in the Department of the Interior, open to criticism at a time when it could ill afford it. When he heard of the appointment of the first director for the new Bureau of Mines, Senator Nathan B. Scott of West Virginia wrote to Taft: "Glad you named Dr. Holmes for head of new Bureau of Mines. Now make some good political appointments that will help. *You have been entirely to* [*sic*] *good*."[114] In the end, Taft did not act politically in the usual meaning of the word. Holmes was the most popular of all the candidates, the best qualified, the most experienced; he had succeeded in convincing Taft that his loyalty to the administration was not a matter of serious question. Holmes was a risk, but a risk Taft in good conscience could hardly afford not to take.

CHAPTER

2

Technology and Politics in the Federal Bureaucracy

THE SAFETY work of the federal government was born in the explosion at Monongah and the fire at Cherry, and the Technologic Branch and the Bureau of Mines were in significant ways reflections of these crisis origins. Due to public apathy and insufficient funds, a number of important coal-mining safety questions received inadequate attention from the branch and the bureau. Falls of roof and coal, recognized by government officials as the number one killer in the industry, were virtually ignored; by fiscal 1915 roof support was being investigated only through a limited cooperative agreement with the state of Illinois. Not until the 1940s would roof-bolting provide an adequate technological solution.[1] Early federal safety investigations also largely ignored the problems of haulage, hoisting, drilling and cutting machinery, conveyors, and electricity.[2] Instead, the bureau inherited from the Technologic Branch an excessive emphasis on explosions, explosives, and rescue. For the fiscal year ending June 30, 1909, the first two categories each absorbed about 40 percent of the expenditures of the Technologic Branch; the third took the remainder.[3] In each of these areas, moreover, the activities of the federal government provoked a surprising amount of controversy.

As it moved into coal-mining safety, the Technologic Branch stepped into a scientific dispute—with immense practical implications—that had been simmering for over a century and boiling for two decades. The great coal-dust debate had its origins early in the nineteenth century. A possible relationship between coal dust and mine explosions was first suggested in England in 1803 and mentioned by Michael Faraday and Sir Charles Lyell in 1835 and A. duSouich in 1855.[4] By the 1880s the controversy focused on the question of whether coal dust could, by itself and without the presence of methane gas, explode and propagate an explosion. The debate had centered on the Continent and evoked no special concern in the United States until the Pocahontas mine in the Flat Top region of Virginia exploded in March 1884. This mine was reputedly free of methane; therefore, when the explosion occurred, the mine superintendent claimed it was caused by fine coal dust. A request by the owners of the mine produced an investigation by the American Institute of Mining Engineers, which concluded that "the explosion was due mainly to dust. . . . we have obtained no direct proof of any past occurrence of firedamp [an explosive mixture of methane and air] sufficient of itself to account for even a slight explosion, and are forced to believe that the explosion was due either to dust alone or to dust quickened by an admixture of firedamp too slight for detection by ordinary means."[5] Apparently, this was the first time in the history of American coal mining that the coal-dust theory was offered as an explanation for a mine explosion.[6]

The general point of view in 1884, however, was not that of the AIME committee but of E. S. Hutchinson, an Englishman who held that dust, by itself and without firedamp, was not explosive. Hutchinson arrived at his conclusion through the back door, arguing that "if the coal-dust theory be true to the extent insisted upon by its most strenuous advocates, every such local occurrence [a blown-out or overcharged shot] in a mine of this description [dry and dusty] should be followed by the more or less complete wreck of the colliery."[7] (A blown-out shot occurs when the force of the explosive is not absorbed by the coal but instead is channeled back out the charged hole. The result—a burst of flame from the hole—may ignite gas or dust in the mine atmosphere.) Europeans continued to investigate the problem during the 1890s, and they disagreed with Hutchinson. British tests from 1890 to 1893 indicated that coal dust alone could cause explosions. The French said coal dust could only intensify an explosion, not be a major agent. Among American engineers, the continental investigations produced some spirited discussions, particularly at the 1894 convention of the AIME, but no commitment to coal dust as an explanation for explosions. From 1895 to 1908, in fact, the *Transactions* of the AIME contain nothing specific on coal-dust explosions. This silence was notable at a time when the danger of coal-dust explosions was rapidly increasing in the United States due to a series of changes in the coal-mining industry. The rapidly rising demand for coal placed a premium on production and encouraged carelessness; recently introduced machinery for undercutting the coal stirred up more dust than hand-mining methods; and greater use after 1898 of the mine-run system of payment for coal increased the likelihood of blown-out shots, as miners ceased to be concerned with bringing out large pieces of coal and began to employ more explosives.[8]

Like the scientists and engineers, operators were divided. Fleming and fellow West Virginia operator Justus Collins took the coal-dust theory more seriously than most, and Fleming was, in fact, privately praised by Holmes for his efforts to prevent dust explosions in his mines in 1909.[9] Other operators were hostile. "Between you and I and the gatepost," wrote William N. Page, "I think our Chief Mine Inspector is a crank on the subject of dust."[10] Seven years earlier Page had discounted the possibility that an explosion in his Red Ash Colliery in West Virginia had been caused by coal dust. Coal dust, said Page, "must necessarily play an important part in all explosions, *when the gas is once ignited*, by adding to the forces involved."[11] A Pennsylvania operator asserted flatly that without a gas explosion, "the coal dust is not likely to take fire. Don't see how it can take fire."[12] It was partly this division and confusion over the facts of mine safety which encouraged operators to turn to the federal government in 1908.

The nation's operators, of course, had no monopoly on confusion, ignorance, or error. The organ of the United Mine Workers went a step beyond the operators and mixed in an element of conspiracy, charging, in effect, that the whole coal-dust theory was an operator fabrication designed to draw attention away from the real problem: ventilation. The *Journal* said: "The 'dust explosion' may be ranked with Bro. Jasper's dogma that the 'sun do move.' It is a safe proposition that where there is no gas in a mine 'dust explosions' do not occur, and that theory is convenient to befog coroner's juries and dull the sword of justice . . . this 'dust explosion' theory was not invented until after prosecutions for disobedience of the mining law were begun."[13] The *Journal*'s attitude was modified slightly in 1905, when its editors admitted the existence of some unknown factor and the need for a scientific investigation. It was obvious even at this time that the *Journal* did not expect scientific research to vindicate the coal-dust theory advocates.[14] One of the most influential mining men in the country, former Ohio inspector Andrew Roy, supported the opponents of the coal-dust theory when he concluded in his serialized and influential book that the coal-dust theory had been "thoughtlessly accepted."[15]

The theory received a tragic stimulus in March 1906, when an explosion in a mine at Courrières, France, killed 1,230 miners, almost four times the largest number killed in a coal-mine explosion in the United States to the present day. The mine had been free of methane or firedamp up to the time of the explosion. The Courrières disaster, said French mine-safety expert J. Taffanel, "has demonstrated in an indisputable manner the reality of the coal-dust danger."[16] American miners remained unconvinced, and less than a month after the explosion in the similarly gas-free Monongah mine, the *Journal* published several letters from miners who refused to accept the coal-dust theory. Perhaps buoyed by its readers, the *Journal* continued to look at the coal-dust theory as an operator facade, just as it continued to trust in Andrew Roy.[17] Opposition to the coal-dust theory was closely related to the instinctive working-class dislike of "experts"—scientific men without practical knowledge of mining conditions. Future president of the mine workers' union T. L. Lewis spoke for thousands of miners: "We have some coal experts running over the country proceeding to tell practical miners about the dangers surrounding [coal dust]. I haven't very much respect for them and I tell them as quickly as I am telling you. They theorize about what the dangers are and how we should protect ourselves. . . . I believe if more was left to ourselves we would be a great deal better off."[18]

Faced with this kind of opposition from a major segment of its audience, it is remarkable that the Technologic Branch made any progress in convincing miners, operators, inspectors, and state legislators that coal dust was dangerous in and of itself. Fortunately, Holmes was a master at getting mileage out of limited funds. Out of the first year's appropriations for mine explosions, Holmes took $3,750, or almost 20 percent, and imported three experts in the coal-mining safety field—Victor Watteyne, inspector general of mines, Belgium; Carl Meissner, councilor for mines, Germany; and Arthur Desborough, H. M. inspector of explosives, England. Their itinerary was arranged to put them in contact with large numbers of coal operators, heads of state operators' associations, and officers of the United Mine Workers. After two months in the United States, the foreign experts prepared a widely publicized report which foreshadowed almost every important facet of the survey's work in coal-mining safety, from permissibility of explosives to (with great emphasis) the explosiveness of coal dust.[19] The foreign safety experts played a critical role in American coal-mining safety. At a time when the federal government was only entering the field, Watteyne, Meissner, and Desborough could speak with authority from years of experience. They influenced operators and miners as well as the direction of the survey's own investigative work.[20]

Once the government scientists had cleared up the confusion in their own minds and published their results, Holmes and his men did enjoy some success in propagating the coal-dust gospel. A. B. Fleming began to employ practices designed to eliminate the danger of coal-dust explosions; the Juanita Coal and Coke Company of Bowie, Colorado, warned its employees of the dangers of coal dust and took precautions. Doubtless other companies did the same, for as early as 1907, the states were beginning to recognize the coal-dust theory in their coal-mining safety legislation, requiring that coal companies take action to reduce their dust accumulations.[21] By far the most opposition to the concept of coal-dust explosions came from the influential *United Mine Workers Journal*, yet even here the scientific and educational efforts of the Technologic Branch began to have impact. Although the *Journal* did not in the next few years acknowledge its error, it did refrain from objecting to the coal-dust theory, and on December 24, 1908, it signaled the end of its resistance with an article which stated that tests by the federal station at Pittsburgh had proved that coal dust in certain stages of dryness and fineness would explode in the presence of an open light or an electric spark.[22] This was an auspicious beginning for the Technologic Branch.

After 1910 bureau scientists focused their investigations on solutions to the coal-dust problem, arriving about 1914 at the conclusion that rock dusting and watering (by spraying with steam, for example) were effective and complementary preventive measures. For limiting explosions, once they occurred, the bureau suggested use of Taffanel barriers, named after their French inventor. The barriers amounted to a series of shelves placed high in the mine, loaded with incombustible rock dust which, when blown down by the advance air wave of an explosion, would cool the flame and prevent the explosion from spreading through the mine.[23] To encourage operators to install rock-dusting equipment and rock-dust barriers, the bureau undertook tests to determine the cost of commercial rock dusting, and, although the process was shown to be significantly cheaper than watering, operators were not receptive. First tried on a large scale in a Colorado mine in 1911, introduction of rock-dusting processes to other American mines was slow.[24] The bureau could only cajole and reason, and such methods proved inadequate; the nation's operators would not act on available scientific and technological information.

The second leg of the federal effort to eliminate mine explosions—the development and distribution of a list of permissible explosives—also met with a lukewarm reception from miners and operators. The idea of "safety" explosives—explosives with shorter and cooler flames—was nothing new to the coal-mining industry, having been applied in Europe for some time before being introduced into the United States in 1901. The essence of the problem was one of labeling. By 1908 numerous manufacturers of explosives were calling their products "safety explosives," appellations which more often than not misrepresented the products and encouraged in the user a misplaced sense of trust. George Otis Smith thought proper labeling was of vital importance to the miner, and in 1908 he suggested federal legislation to regulate the manufacture, use, and sale of explosives.[25] Holmes, too, was impressed by the urgency of the problem, especially the need to provide miners with information regarding the quality of explosives and the quantity of explosives necessary for safe use under different mining conditions.[26] Under Holmes, the Technologic Branch conducted research on explosives, established safety standards, tested explosives submitted by manufacturers, and succeeded in persuading manufacturers to develop newer, safer explosives. Explosives testing began almost immediately after passage of the 1908 appropriation act, and by May 1909 the Technologic Branch had published its first list of "permissibles."[27] Opposition to the new explosives was particularly strong among miners, usually because of some impact, direct or indirect, which the safety explosives had on miner compensation. Since miners often had to pay for the explosives they used, they were sensitive to the higher costs of permissibles. In 1909, for example, the opposition of Ohio miners to a piece of legislation which would have prohibited the use of the traditional black powder was based primarily on the anticipated higher cost of substitute explosives. Miners in Pennsylvania's District 5 staged a major strike when they calculated that the safety explosives which the operators had chosen would reduce earnings by shattering the coal unnecessarily.[28] For whatever reason, permissibles did not exactly sweep the field, and as late as 1922 they accounted for only 18 percent of the explosives in use. Experience did not bear out one optimistic prediction that the mere existence of a permissible list would have a "moral effect" almost as great as statutory law.[29]

The federal government carried out training, investigation, and rescue functions through its mine-safety stations. The first station was established in 1908 at Urbana, Illinois, and by 1911 there were twelve

—six stationary and six mobile—all in coal-mining areas. The mobile stations were maintained in railroad cars, which moved from one mining camp to another within fixed districts, training miners in first aid and rescue techniques, providing general instruction about mine safety, and, if necessary, participating in actual rescue work. From the beginning, the training and investigation functions were primary, although by June 1914 the bureau had rescued approximately one hundred trapped miners. The stations were under J. W. Paul, recently acquired from the state of West Virginia, where he had been chief mine inspector.[30] In 1910 Holmes, convinced of the importance of the program, sent Paul and Carl Scholz, consulting mining engineer, to five European countries to inspect their facilities to determine how to equip the cars and stations; in 1913 the bureau director recommended that all the stations be converted to cars (the miners who were not close to the stations were not coming, he said) and concluded: "I know of no other investment the Government could make that would do as much as this for safety in mining and good citizenship among the 2,000,000 men in this industry."[31]

As a group, miners and operators received these services gratefully. Operators, particularly, could see no reason to resist free training for their employees; Oklahoma operators even agreed to furnish free room, light, heat, and rent if the government would send one of its cars. As long as the Technologic Branch and the bureau asserted no legal right to investigate mine disasters and their causes, they faced little opposition from operators in this area, either.[32] And when money savings were involved, operators could be generous in their thanks. A Gebo, Wyoming, operator wrote in reference to the bureau's rescue work following a January 1917 mine fire: "I am not proficient enough in the use of the language to express my own feelings towards the Bureau of Mines. All of the equipment we had installed under your regulations had to be used. Without this apparatus we quite likely would have had to seal up the entire mine tight."[33] Bureau records contain hundreds of requests from operators and local and district unions that government rescue cars visit their mines.

The generally favorable response to the bureau training program was in part the result of the efforts of W. D. Ryan. A former operator, Ryan resigned in 1913 from the presidency of the Southwestern Interstate Coal Operators' Association and traveled the nation for the bureau as its mine safety commissioner, attending miners' conventions and first-aid meets and calling informal gatherings in an effort to stimulate miner and operator interest in first aid and rescue. Much of Ryan's energy was directed toward institutionalizing interest in the form of meets and contests, and in this and in simply arousing interest, he was remarkably successful. Ryan arranged for first-aid and rescue organizations in Arkansas, Oklahoma, Kansas, Iowa, and other states.[34] Favorable evaluations of Ryan's work came from Holmes, who appreciated Ryan's ability to secure information from the miner's viewpoint, and from Ryan himself, who in 1916 wrote with pride of the escalating interest in his work. Despite his operator origins, Ryan established close relationships with miners' unions in the Southwest and Midwest, and to him must go much of the credit for whatever success the bureau first-aid and rescue programs enjoyed in these coal-mining regions.[35]

Nonetheless, this program did not escape its share of criticism. Federal rescuers were scored for their lack of courage, for excessive interest in saving property rather than lives, and for arriving when everyone was dead.[36] The Cherry mine fire focused the attacks of critics. The United Mine Workers led the foray, charging that survey priorities were at fault and that more attention should be paid to preventing disasters than to rescue efforts. Criticism of priorities was mixed with the *Journal*'s anti-expert attitude here, too. The Geological Survey, charged the *Journal*, "seemed to be getting further away from a solution of this matter, rather than closer to it. There was a whole army of experts at the terrible affair at Cherry, Illinois. . . . They had all the necessary appliances for rescue work, and yet not a single life was saved from the burning mine."[37] The *Journal* had told only part of the story. In fact, the Geological Survey team from Urbana was notified late of the Cherry disaster and the Pittsburgh rescue station learned of it only through the newspapers. *Outlook* magazine concluded that the real lesson of Cherry was the need for all companies to have a rescue corps at their mines; it also suggested that officers of mining companies be required to telegraph at once the appropriate government rescue stations and be held responsible for failure to do so.[38] Along the same lines, James Callbreath, secretary of the American Mining Congress, used the failures of the Geological Survey at Cherry to support his bid for a bureau of mines, insisting that the reason the survey was not notified immediately of the Cherry disaster was that few people associated the survey with rescue work. A bureau of mines, said Callbreath, was the only solution: "The saving of the lives of miners is not a task for geology, but for eminent scientific mining engineers."[39] Finally, the survey gained knowledge of its opposition from the criticism it faced after Cherry. Holmes knew that his men on the rescue cars would encounter negativism from miners who "will not care very much about theories."[40]

Federal progress in coal-mining safety faced two major political challenges after 1910. The first came from Congress, whose initial enthusiasm for the work of the Bureau of Mines was now tempered. From 1911 through 1913, the bureau was hard-pressed to maintain even a no-growth budget. All forms of bureau work—safety, conservation, mineral development—were affected by congresses which, while not hostile to the bureau, neither viewed it as comparable to the Department of Agriculture nor saw the research needs of mine operators as comparable to those of farmers. The second challenge was the more serious, for it had the strong support of Holmes and his successor as director, Van H. Manning. Under their leadership, the bureau shook free of the strong influence of Monongah and Cherry and concentrated its energies increasingly on western metals and conservation. In the process, coal-mining safety became only another aspect of the bureau's work; no longer was it the bureau's raison d'être.

The roadblocks to adequate funding lay not in the executive branch, where both Taft and Wilson favored mine-safety investigations, but in congresses which were no longer much interested in mining or mining safety. "Membership on the Agriculture Committee," said the *Pittsburgh Gazette Times*, "is considered one of the honors, while the Committee on Mines and Mining goes begging for a chairman or for membership."[41] A Seattle attorney, Maurice D. Leehey, compared congressional attitude toward the mining industry and agriculture: the 1914 appropriations for the Bureau of Mines, wrote Leehey, "just equaled the amount appropriated by Congress for investigations in the treatment of hog cholera. In other words, the entire mining industry received the same consideration as one single disease affecting one farm product."[42] Thomas J. Walsh, then chairman of the Senate Committee on Mines and Mining, replied: "Your comparison of the appropriation made for the support of the Bureau of Mines with the liberality exhibited by Congress touching hog cholera, illustrates strongly the niggardliness and indifference with which the mining industry is treated by Congress." But even Walsh found it difficult to keep informed of the work of the bureau as other demands on his time became more numerous and compelling.[43]

Because of a Taft campaign for economy and efficiency and a recession in 1913, the bureau had the greatest difficulty with Congress in its early years. Total bureau appropriations fell slightly from fiscal year 1911 to fiscal 1912, in spite of assistance from a UMWA legislative committee and from Secretary Ballinger, who supported an increase in appropriations for new rescue cars and stations. Without it, Ballinger concluded, "the progress of the work will be seriously handicapped."[44] In the bureau's *First Annual Report* Holmes noted ominously that funds available for investigations of the causes of accidents and explosions "instead of increasing, in accord with the experience of other bureaus, are actually less to-day than they were for the technologic work under the Geological Survey three years ago."[45] In fiscal 1913, these investigative funds were actually reduced, from $347,000 to $327,000. While total bureau funding doubled from 1911 to 1917, when the bureau became a million-dollar organization, appropriations for scientific investigations of the causes of mining accidents increased only slightly in that same period, with the entire increase coming in funding for decentralized investigations at mining experiment stations in *metal* mining areas. Funding for scientific work in coal-mining safety did not increase over the entire ten-year period from 1910 to 1920. Although the research needs of the mining industry were often compared with those of agriculture, in actual funding mining was a distant second. For fiscal 1913, the federal government spent almost $28,000,000 on agriculture and only $1,967,000 on mining, though mines produced almost half the yearly value of agricultural products. In 1912 there were fifty federal agricultural experiment stations; mining had fewer than ten.[46] That same year, appropriations for operating the bureau's mine-rescue cars were exhausted by March 1, 1913, and Holmes's attempt to secure an emergency deficiency appropriation from Congress failed. Operation of the cars had to be discontinued temporarily.

Appropriation difficulties delayed investigations of roof falls, improvement of rescue equipment, and examination of fuel and mineral waste. Construction of an experimental mine for the bureau took three years when, according to Holmes, it could have been done in one with adequate funding. Because it was primarily an educational institution, the bureau suffered greatly when its educational efforts were hamstrung by appropriation deficiencies. In 1912 the bureau's publication funds were cut to 40 percent of their 1911 level, and in 1914 the secretary of the interior, Franklin K. Lane, had to come to the rescue of the bureau when Congress threatened to eliminate all funds for attendance of bureau employees at association, institute, and society meetings. Lane effectively argued that miners could be most readily interested in safety problems through in-person lectures and demonstrations at appropriate meetings, and that an absolute prohibition of attendance at such meetings would cripple the educational work of the bureau. Financial problems endangered the progress of the Technologic Branch and the bureau with mine inspectors and owners and shook the confidence of the miners in the good faith of the national government. By 1914 Holmes was forced to conclude that "the claim frequently set forth of late by the miners, mine owners, and inspectors that the entire mine-safety movement is being held back by the lagging of the Government's investigations is unfortunately a true and reasonable claim."[47]

Always the political activist, Holmes took up where he had left off with the Geological Survey. He worked hard to counter congressional negativism, courting members of Congress and particularly the members of the committees on Mines and Mining. In addition to ordinary persuasive techniques, Holmes and Manning offered entertainment in the form of the mine-safety demonstration. Yet even Holmes's considerable abilities in public relations and politics proved ineffectual. The First National Mine Safety Demonstration, held in Pittsburgh in October 1911, attracted the president but apparently only one congressman, George White of Ohio, and by early 1912 Holmes had managed to get only two members of the House Committee on Mines and Mining—White and William B. Wilson—to the Pittsburgh experiment station to see the explosion work carried on there.[48] The forces inducing congressional apathy toward mining and coal mining were simply too strong to be countered from within the bureaucracy. Except for 1913, when 233 died at Dawson, New Mexico, and 181 at Eccles, West Virginia, there were no disasters like Monongah in the decade after 1910, and the general public and the media, inured to small mine disasters, were no longer pressing for reform. The coal-mine operators might have conquered this indifference with effective lobbying, but they were unable to develop an organization which could speak with authority for the entire coal industry.[49]

Their failure left the field to the American Mining Congress, traditionally and still essentially representative of western metal-mining interests. Working with Holmes, these interests were responsible for the Foster Act of 1913, which clarified and replaced the 1910 legislation creating the bureau, and the Kern-Foster Act of 1915, which extended the experimentation, rescue, and educational programs of the bureau from coal mining into metal mining. Along with World War I, this legislation considerably reduced the bureau's concentration in coal-mining safety. Proponents of the Foster bill were unanimous in their claims that its enactment would not extend the scope of the work of the Bureau of Mines; its function was only to state that scope more clearly, in accordance with the original intent of Congress.[50] But supporters of the bill also realized that clarification of the 1910 act would result in major changes in the bureau. Appearing before the Senate Committee on Mines and Mining, a circumspect Holmes emphasized the need to move the bureau into two areas: metal-mining safety and conservation (more precisely, the prevention of waste in mining operations). On the first point, Holmes was careful not to threaten metal mine operators: "We have had . . . in the coal mining regions recommendations by various people who are interested purely in safety that we should submit to the State legislatures recommendations compelling operators to do certain things in behalf of safety which would be practically impossible for the operators to do, because it involves so large an increase in expenditures that there are no possibilities in the industry that would enable them to do these things. I mention this simply to show that the question of economy and efficiency in mining must be considered along with the question of safety; as otherwise we might recommend absolutely impossible things."[51]

In his goals for the bureau and in his cautious approach, Holmes had the support of John Mitchell. In a letter which Holmes called the "most important recent contribution to the mining situation in the United States," Mitchell wrote Martin Foster, head of the House Committee on Mines and Mining:

> As one of the advocates of the establishment of the Bureau of Mines, I have watched with interest the good beginning it has made; but I have noted with regret the inadequacy of its facilities for taking up even the matter of coal mine accidents in a manner commensurate with the urgency and importance of that subject; also the omission from its work of investigations looking to the general up-building of the industry. . . .
>
> If I may be permitted to speak for the coal miners, I can express for them the desire that whatever may be done in behalf of their safety should also be done in behalf of the safety and welfare of the men who work in the quarries, metallurgical plants, and the metal mines, where the loss of life and health is but little, if any, less than in the coal mines. If I may be permitted to speak for all classes of miners, I will say for them that they are unwilling to have the work of this Bureau limited to endeavors in behalf of their own safety and welfare; but desire its extension with a view to the upbuilding of the

> industry of which they are but a part. The men who labor underground, with all its attendant hazards, no less than other classes of our citizens, have at heart the permanent welfare of this country. They, more clearly than any one else, realize the deplorable extent to which the gas, coal and other essential resources of this country are being wasted. And they believe that the first move to be made in determining a feasible plan for solving the difficult problem of reducing this waste, is to provide for a thorough investigation into the facts of the situation by the Bureau of Mines, and a publication of the results.[52]

Mitchell's letter is of interest for several reasons. There is some question, especially in light of his close relationship with the National Civic Federation (NCF), whether Mitchell could speak with any authority for the nation's coal miners. In fact, the statement is a good example, not of the rhetoric of the United Mine Workers, but of the NCF's attempts to paper over the differences between capital and labor. Finally, the statement begins with safety but concludes by emphasizing industrial development and conservation, hardly the priorities of coal and metal miners.

There is no evidence that the United Mine Workers supported the Foster legislation. Rather, its genesis was corporate and sectional. By far the most active proponents of the Foster bill were western metal-mining interests, anxious to see the Bureau of Mines invest an increasing portion of its time and energy in the western metals industry. James Callbreath of the American Mining Congress summarized the general feelings of this group when he suggested that appropriations for the Bureau of Mines were totally inadequate. It was, he said, "a case of 'All for the white man and none for the nigger.' "[53] Although western metal mine operators agreed with the need for more attention to metal-mining safety and waste, their basic needs were more utilitarian. They wanted the bureau to move rapidly into the metallurgical treatment of low-grade ores.[54]

Mining engineers, apprehensive that their own careers and status would be directly affected by competition in commercial work from the federal government, were the bill's most serious opponents. One engineer wrote that he did not think it "likely that the bureau employees will ever be able, in the casual study they will necessarily have to make of different metallurgical and mining questions, to attain the expert knowledge of these questions possessed by the engineers who have devoted years to their study."[55] The *Engineering and Mining Journal*, organ of the Mining and Metallurgical Society, denounced federal work in metallurgical testing, and the president of that society,

J. Parker Channing, wrote to Foster in opposition to his bill. "Broadly speaking," said Channing, "I believe that the Bureau of Mines should not be permitted, either directly or indirectly, to go into commercial work, that it should devote itself to looking out for the health and safety of persons employed in that industry."[56] Engineering opposition to the legislation was often phrased in terms of the impact the bureau's enlarged activities would have on private enterprise.

Broad enough for all its advocates, the Foster bill became law in February 1913. Its logical successor was the Kern-Foster Act, also symbolic of the influence of western metal mine operators in bureau policy and indicative of the position of coal-mining safety in the dynamics of the bureau at this time. Approved in March 1915, it called for the establishment and maintenance in unnamed mining regions of ten mining experiment stations and seven mining safety stations (movable or stationary) and represented the consolidation of hundreds of requests from all parts of the nation for experiment and rescue stations. Although the bill's strongest support came from the metal-mining industry, which felt that the bill would go far toward redressing the imbalance in the bureau toward coal mining, passage of Kern-Foster also was favored by the United Mine Workers, the American Federation of Labor (AF of L) and the American Association of State Geologists. A committee of the UMWA and the AF of L was particularly helpful, and William Green was active on the bill's behalf. Holmes used all his influence in behalf of a bill he felt was, "other than the establishment of the Bureau itself, . . . the most important mining legislation proposed for some time."[57] The year before, lack of funds for the mine rescue cars had forced the bureau to discontinue operation of some cars, and Holmes anticipated that passage of Kern-Foster would place the program on firmer foundations. Although Kern-Foster did not itself provide funds for the new stations, the act required that three of the mining safety and mining experiment stations be established in the current year. Writing to state geologists and others from Fort Bayard, New Mexico, where he was undergoing treatment for tuberculosis, Holmes promised experiment or rescue stations to particular states, apparently in return for political support. Holmes fully admitted that he was not authorized to do so, but he promised stations to cities in Montana, Minnesota, and Kansas, and to the University of Alabama. When Holmes died soon after passage of the act, Van H. Manning, acting director, did not know all the states in which Holmes had proposed to locate the stations. He wrote to the president of the

Association of State Geologists, F. W. DeWolf of Urbana, that the secretary of the interior had already turned down one location "which had been rather definitely promised by Doctor Holmes."[58] Clearly Holmes was willing to go to some lengths to assure passage of the Kern-Foster bill.

These events brought early changes in bureau priorities. By 1913 the old survey problems of coal-mining safety and fuel use, while still significant, were joined by a third of equal status: safety (and "efficiency," to use Holmes's euphemism) in metal-mining operations. The years 1914 and 1915 witnessed the beginnings of a decline of health and safety problems in the bureau's priorities and a commensurate increase in interest in a variety of areas, including smelter smoke, the production of radium, and the manufacture of gasoline, benzol, and toluol from petroleum. The work of the bureau's Petroleum Division was becoming increasingly important, as had that of the Metallurgical Division, which made only limited safety investigations. The trend away from coal-mining safety is illustrated, too, by the bureau's cooperative agreements (which were essentially cost-sharing arrangements). In fiscal 1916, only one of nine official cooperative agreements involved coal-mining safety; most dealt with metal-mining safety.[59] By the end of fiscal 1919, all but one of the seven mine-rescue cars and all but four of the ten mine-experiment stations authorized by the Kern-Foster Act had gone into service. Nearly all the new cars and stations served metal-mining areas, and no safety work was done at any of the experiment stations. The act itself is important as an indication of the status of coal-mining safety at the time of its passage; its implementation—largely in the metal-mining areas—is a reflection of the powerful competition of metal-mining safety and technology for bureau funds in the second decade of the century.[60]

World War I accelerated this relative decline in interest in coal-mining safety. In the long run the war, by encouraging a national, collective response to problem-solving, may also have encouraged national action in health and safety; in the short run, however, the effect was to promote development of the Bureau of Mines in areas totally divorced from mining and mine safety. Bureau resources devoted to coal-mining safety remained fixed; those in other fields grew rapidly. Specifically, the bureau became deeply involved in investigations of the use of gases in warfare, production of nitrates and new alloys, conservation of fuel, and the possibilities of developing domestic supplies of metals such as nickel and manganese. Another war-related deterrent to safety investigations was the shortage of engineers for field investigations. Although the war involved the bureau in explosives regulation, the goal was national security, not safer mines.[61]

In his five years with the bureau, Joseph A. Holmes had succeeded in removing the inhibitions of the 1910 organic act and in implementing the new Foster legislation. The bureau continued, however, to be restricted in another sense; with the exception of a short period during the World War, it remained an educational agency without regulatory powers of any kind, legal or administrative. "Hence," said the bureau's second director, Van H. Manning, "as the bureau can not forbid nor compel, it recommends and advises. It appeals to reason, not to fear. Its campaign for greater safety is essentially a campaign of education."[62] As an educational institution, the bureau developed and used a number of teaching and publicity techniques. It published *Circulars* for miners and *Bulletins* and *Technical Papers* for operators and professionals in the mining-safety field; it held national mine-safety demonstrations and, in conjunction with other public and private agencies, sponsored first-aid and rescue meets around the country; its representatives spoke at meetings of miners, inspectors, engineers, and operators; in its most public and widely known activity, the bureau, through its stationary and movable cars and stations, tried to reach miners and operators to instruct them in the proper techniques and equipment of rescue and first aid.

Mine safety was not always the most newsworthy of topics, and good national publicity was difficult to obtain. The Illinois Mine Rescue Commission, for example, was known to have paid for coverage.[63] In this setting, the national mine-safety demonstration became vital to the bureau's educational effort. The idea appears to have come from H. M. Wilson, engineer-in-charge of the Bureau of Mines Experiment Station at Pittsburgh; after conferences with congressmen, senators, union officials, and state mine inspectors, the First National Mine Safety Demonstration was held in October 1911 and was attended by President William Taft and thousands of miners. The demonstration included a first-aid exhibit, a general demonstration of various means of making mines safe, and, perhaps most important, was the occasion for the first public use of the now famous slogan "Safety-First," developed several years earlier for the Illinois Steel Company.[64] The bureau held another national exposition in 1913, took part in the preparation and conduct of a National Safety-First Exposition in February 1916, and sponsored hundreds of mine-rescue and first-aid contests in the years from 1910 to 1920. Creating as well as reflecting the current concern with industrial safety, the bureau's educational efforts led a national safety figure to term the agency "a pioneer of concerted training in first aid in the industrial world."[65]

Of the three types of bureau publications, perhaps the most important were the *Circulars*, distributed free to some 13,000 mine officials—fire bosses, shot-firers, superintendents. Roughly equivalent to the Farmers' *Bulletins* issued by the Department of Agriculture, the *Circulars* were simply written so that these officials, who seldom had more education than the miners working under them, could grasp essential new knowledge in mine safety and pass it on to their subordinates. In fiscal year 1912 the bureau distributed more than 500,000 total publications, including about 352,000 *Circulars*.[66] While generally well received, bureau publications were sometimes denounced for persistence in advocating a coal-dust theory of mine explosions and for overemphasizing the control a miner had over his own safety. One particularly virulent Indiana miner, T. James, suggested that the advice offered in *Miners' Circular No. 11*, if followed, "would be undoubtedly the best step yet taken in the evolution of the mining industry from a slaughter house to an under-ground paradise, were it not for the fact that the men it directly appeals to, (the miner and the foreman) are undoubtedly unable to follow it." Referring to statistics presented in the *Circular*, James believed: "Ye Gods! Just look how magnaimous [*sic*] the management must be. Not one life lost . . . by any fault of the owners."[67] Perhaps in response to these and similar comments, Holmes established a policy of sending the *Circulars* to leading mining men for suggestions and comments. John Mitchell, John White (when president of the UMWA), and William Green were among those who offered advice.[68]

For its rescue and first-aid program, the bureau had expectations that went beyond education of miners. In theory, at least, this work was only pump-priming; private businesses would see the obvious benefits in maintaining their own equipment and instructors and in training their own miners. Holmes, Manning, Ryan, George Rice (the bureau's chief mining engineer), and some coal operators accepted versions of the theory, which was first stated publicly in the 1912 *Annual Report* of the director. By 1918 the formula had been refined by Holmes and Manning so that it now called for each mining operation to maintain at least 10 percent of its employees in rescue and first aid, with a station at the mine or managed cooperatively with adjacent mines.[69] Although the idea of a self-liquidating federal program has the ring of ideology, government and business backing for the concept stemmed from a pragmatic base. While Holmes may have felt strongly that private industry could do the job more efficiently, his strongest statements on the subject came during appropriation hearings in 1912, 1913, and 1914, when the bureau was locked in budgetary struggles with Congress. In 1913, for example, Holmes told the Appropriations Committee: "I am doing everything, and the other officials of the bureau are doing everything they can, to unload this work as rapidly as we can upon the state and the private operators."[70] In short, the pump-priming theory was essentially a response to short-term fiscal necessity, enunciated to prove groundless congressional fears of burgeoning spending in a new government bureau by showing that a good deal of the financing was temporary.[71] Holmes was also flexible enough to see the advantages of a rescue system financed by the states rather than the private sector, a system that promised to be operationally inferior but more consistent with bureau financial limitations. Manning and a number of operators, on the other hand, supported pump-priming and an eventual all-private system on the grounds that adequate rescue service required the involvement of individual operators. The nature of rescue work lent some credence to this view, for one of the greatest liabilities of a federal or state system, even one using movable cars, was the delay involved in reaching a disaster. Illinois operator A. J. Moorshead, president of the Madison Coal Corporation, wrote: "We have believed that the very best work can only be accomplished by every mine being organized in such a way that it can take care of itself."[72]

Private participation was not lacking. The Cottrell report of 1915, a major in-house attempt to assess bureau progress in a number of areas, was laudatory of bureau gains in this field: "Whereas there was not a single mine rescue station or car in the United States when the bureau adopted these means of training and rescue," said the report, "mining companies have now established 76 mine rescue stations at which there are 1200 sets of artificial breathing apparatus besides the auxiliary equipment for fire fighting."[73] While this would seem to indicate substantial progress, the self-professed bureau goal—10 percent of mine employees well trained in first aid and mine rescue—was not approached even in the largest (and generally most progressive) coal mines. A 1916 bureau survey of the nation's coal companies employing more than two hundred persons reveals that only in exceptional cases did the total number of mine-rescue and first-aid personnel employed at a particular company reach 10 percent, and more commonly

TABLE 2

Number of Rescue and First-Aid Personnel and Medical Facilities in Private Coal-Mining Companies of More Than Two Hundred Employees, by State, in 1916

State	Number of mines: Total	Number of mines: With hospital service	Mine-rescue men per 1,000 employees	First-aid men per 1,000 employees	Surgeons per 1,000 employees
Alabama	8	5	13	26	4
Colorado	5	5	44	41	5
Illinois	30	18	17	21	2
Iowa	1	0	17	0	0
Indiana	4	1	12	9	2
Kentucky	10	4	14	36	4
Maryland	2	1	0	7	1
Montana	2	1	29	43	3
Oklahoma	2	2	5	2	5
Ohio	7	4	3	3	2
Pennsylvania anthracite	34	33	14	32	1
Pennsylvania bituminous	43	29	8	28	2
Tennessee	2	0	6	19	6
Texas	3	3	0	0	2
Virginia	5	5	5	6	2
Washington	4	3	24	46	3
West Virginia	23	16	4	7	4
Wyoming	6	6	31	42	3

5 percent or less of the employees had such training. Private facilities were particularly sparse in Illinois where the state was heavily involved. (See Tables 2 and 3.)[74]

By the time Manning took control of the bureau, the theory and practice of state action in coal-mine safety education had severely undermined the possibilities of a strong private system. The challenge was led by Illinois, which had previous difficulties with the bureau over rescue services and which by 1914 had three of its own rescue cars. "In Illinois," said Holmes, "the fact that the State is doing this work on so extensive a scale appears to have discouraged private mining companies in that State from doing it."[75] When Holmes made this statement in 1914, few Illinois coal-mining companies maintained private rescue cars. After 1917 the profit motive would enter the picture in the form of an insurance group called the Associated Companies, and private action would receive a strong stimulus.[76] The bureau was partially responsible for stimulating the birth and growth of state systems. Its staff prepared the legislation creating the Illinois system and helped operators and the state of Tennessee work out a cooperative rescue station agreement for the Jellico field in 1913. The

TABLE 3

Number of Rescue and First-Aid Personnel and Medical Facilities in Private Coal-Mining Companies of More Than Two Hundred Employees, by Size of Company, in 1916

Size of company by number of employees	Number of mines Total	With hospital service	Mine-rescue men per 100 employees	First-aid men per 100 employees	Surgeons per 100 employees
200–225	13	8	13	41	6
226–300	42	31	14	22	4
301–325	3	3	4	55	2
326–350	6	3	22	26	3
351–400	15	8	8	17	4
401–450	6	5	12	16	3
451–500	13	8	18	27	3
501–550	8	5	8	49	3
551–600	9	5	12	12	2
601–650	7	5	6	20	2
651–750	15	12	10	21	1
751–850	5	4	15	21	1
851–950	4	3	16	34	3
951–1,000	1	0	12	0	1
1,001–2,000	24	19	13	19	2
2,001–3,000	7	6	12	23	1
3,001–4,000	3	3	3	19	0
4,001–5,000	4	2	9	29	3
5,001–10,000	3	3	10	13	0
10,001–15,000	2	2	17	30	0
15,001–20,000	No companies in this size category.				
20,001–up	1	1	22	56	1

ubiquitous W. D. Ryan worked closely with operators and miners in a successful effort to obtain funds from the Kansas legislature and governor for a mine-rescue station in Pittsburg, Kansas. When George Rice strongly objected to these bureau activities on the grounds that they militated against private systems, Manning could only reply: "It seems to me that the only way in which these stations can be organized and maintained is by law, until the operators have been educated and will give their support to the cause without duress."[77]

The mine-rescue cars and stations were the most public aspect of the bureau's work to contemporaries, and their importance was stressed by the bureau directors, but beyond these essentially educational efforts, the bureau had other functions, practical and scientific, which deserve mention. In the practical realm, the bureau was a testing agency, essentially conducting two types of tests. First, it examined samples of coal dust and mine gas received from mining companies, providing technical data for company engineers and an occasional recommendation if the dust proved particularly dangerous. Second, the bureau, having established noncompulsory standards for coal-mining safety products, tested explosives, motors, switches, portable electric lamps, storage battery locomotives, mine-lamp cords, flash lamps, flame safety lamps, gas detectors, and coal-cutting apparatus to determine if particular products conformed to its standards. The bureau also contributed to future coal-mining safety through scientific investigations. Particularly consequential were its ventures into rock-dusting and its investigations of explosives, the latter continued from the Technologic Branch.[78]

The bureau was aided in its work by effective internal administration —characterized by efficient organization and a high level of staff continuity and employee morale—and by a network of cooperative relationships with public and private organizations outside the bureau. Although a number of these cooperative relationships were productive of little more than administrative friction, others were important to the coal-mining safety effort. As a group, they are indicative of Holmes's and Manning's attempts to employ all available resources in the safety movement and of the bureau's need to compensate for its statutory deficiencies.

The Bureau of Mines was organized around five divisions: Administrative, Mining, Mechanical, Chemical, and Mineral Technology. Head of the Administrative Division was Holmes, with his assistant director Van H. Manning. Head of the Mining Division, under which most of the mine-accident work was done, was George S. Rice, chief mining engineer. Explosives and mine gases were analyzed in the Chemical Division, supervised by chief chemist G. A. Hulett. Coal-dust explosion investigations were carried on at the bureau's Pittsburgh Station under the direction (through most of the period) of H. M. Wilson. Wilson's efforts and the mine-rescue and first-aid work of J. W. Paul (until 1915) came under Rice and constituted the major portion of the activities of the Mining Division. There were a few efforts to change the structure of the bureau. A 1915 report suggested reorganization according to mineral products (i.e., coal-mining division, metal-mining division), but it was vetoed by Rice, who claimed the result would be inefficient, since there were so many similarities between coal mining and metal mining in terms of gases, ventilation, timbering, and explosives. Rice also objected to a suggestion made in another report by chief chemist F. G. Cottrell (later, briefly director of the bureau), that all the safety work of the bureau be handled by one man designated as safety engineer. Perhaps Rice resisted the proposal because it would have meant an increase in the influence of H. M. Wilson and a decrease in his own responsibilities. Whatever the reason, the bureau was not reorganized internally either by mineral or by problem (e.g., safety), and it does not appear to have suffered for maintaining its original organization.[79]

Holmes, and later Manning, maintained close supervision of the various bureau divisions, and perhaps this personal approach was responsible for what Cottrell termed "the rather striking spirit of personal loyalty throughout the Bureau."[80] Bureau records reveal only one major conflict between bureau personnel in the ten years after 1910, and that was handled effectively by Holmes.[81] The small turnover in high bureau positions is also indicative of the harmony produced by Holmes's leadership. Manning and Rice, for example, were with the bureau until at least 1920; the ambitious H. M. Wilson stayed until 1915; F. G. Cottrell and O. P. Hood joined the bureau in 1912 and remained through 1920.

Of the cooperative relationships, those of a legal nature were entirely with state agencies. In 1914 the bureau entered into an agreement with the Utah Industrial Commission whereby the two parties jointly employed a mining engineer and Utah furnished clerical assistance, office space, supplies, and assistants. Not until 1919, however, did this cooperative agreement involve coal-mining safety investigations. By then bureau cooperative agreements with states and state institutions numbered eighteen, most of them with universities. Eight of these dealt with mine safety, and three—Utah, Colorado, and Illinois agreements—made provision for coal-mining safety work of some kind. One—an agreement between the bureau and the Colorado School of Mines to sponsor a Holmes Chair of Safety and Efficiency Engineering—was one of many tributes to the bureau's first permanent director.[82] The most potentially significant of the legal state-bureau agreements was that with the Illinois State Geological Survey and the Mining Engineering Department of the University of Illinois. Initiated under the Technologic Branch of the survey to provide a comprehensive examination of coal-mining conditions in Illinois in regard to waste and safety, in 1914 the bureau allocated $7,800 and Illinois state $9,500 to this work. Although this particular agreement was renewed in regular three-year periods throughout the decade, there was general disappointment on the part of participants with its accomplishments and its direction. Illinois's objections centered on the tendency of the cooperative project to overemphasize conservation and efficiency at the expense of safety. George Rice saw few of the investigations under this cooperative agreement as productive, and Manning was also dissatisfied.[83]

Unlike its attitude toward relationships with the states, the bureau tended to be suspicious of the safety work of other government agencies.[84] With the exception of its agreement with the Public Health Service and its willingness to cooperate with other governmental agencies in national safety expositions, bureau prerogatives in the safety area were carefully upheld. While the origin of this defensive attitude is not clear, in 1913 the problem became acute, at least to bureau personnel, when a rather minor matter created a large disturbance. At that time the United States Weather Bureau, without consulting the Bureau of Mines and operating under the assumption that the mine operators in a Pennsylvania region should be informed of sudden changes in barometric pressure in order to prevent dust explosions, began sending out "miners' forecasts." Enraged, Holmes wrote to the chief of the Weather Bureau: "If . . . this is a general policy entered into by the Weather Bureau in all coal mining States, it is simply a more emphatic illustration of an unfortunate practice in parts of Government work, of each bureau proceeding with plans of its own without exhibiting any spirit of cooperation with other bureaus interested in or conducting kindred investigations."[85] The Weather Bureau eventually stopped issuing the forecasts, but as late as 1916 one bureau official recalled the incident and used it as a warning "against getting mixed up in any way with other Government bureaus or departments."[86]

The threat of a possible new government agency, rather than the actions of an existing one, provided an even greater challenge for the bureau. Here the chain of events began in 1913, when the first bill calling for the creation of a bureau of labor safety was introduced in the House of Representatives. The idea was to coordinate the government's safety work under one bureau and move it to the Department of Labor. Holmes's reaction was predictable. The Lewis bill, he telegrammed Robert N. Page of the House, "will cripple the work the Bureau Mines by taking away all its safety work."[87] "This sort of duplication," Holmes wrote to Foster, "is at the present time the curse of the Government service; and there are several bureaus in the Government service that are pushing their work with an absolute disregard as to whether or not the field is already occupied by other bureaus."[88] In 1914 Holmes was successful in attaching to the relevant bills for a bureau of labor safety an amendment protecting the safety work of the Bureau of Mines from transfer to any new bureau, and in

1916, when the subject again threatened the Bureau of Mines, Manning secured the defeat of another bill through mobilization of coal and metal mine operators, the United Mine Workers, state geologists, and other interested parties.[89]

More important to the bureau's coal-mining safety program than cooperative relationships with other federal agencies were those with private organizations, though present here, too, were problems of jurisdiction. The process by which such relationships evolved is clearly illustrated in the case of the National Fire Protection Association (NFPA). Correspondence between the NFPA and the bureau was initiated as early as 1911, the year H. M. Wilson of the bureau addressed the NFPA annual convention on the subject of mine fires. During 1912 R. Y. Williams, engineer in charge of bureau work at Urbana, Illinois, developed a plan for cooperative investigations between the bureau and the NFPA. George S. Rice, for one, was suspicious: "I hesitate to express my opinion," he wrote to H. M. Wilson, "not knowing what led up to it, and to what extent the Bureau might or might not be limited or hampered in its work, or on the other hand, assisted."[90] At the time of the May 1912 annual meeting of the NFPA, a Committee on Mine Fires was in existence, including Wilson and Williams in its membership. The committee report for that year was essentially an attempt to delineate the respective functions of the Bureau of Mines and the NFPA as they concerned prevention of mine fires; the committee decided, for example, that while the bureau should collect, tabulate, and publish statistical data on mine accidents, the NFPA should make recommendations as to the nature of the inquiry and the type of data desired. The bureau was assigned fire-fighting methods, the committee was to consider the size and construction of water pipes and the relative efficiency of various pumps, nozzles, and fire extinguishers. In later years, the NFPA published a newsletter in which the public was informed of bureau publications relating to mine-fire prevention.[91]

The NFPA was a fire-safety organization. One small facet of its business was mine fires, and only in that one area did it come into contact with the Bureau of Mines. As an organization of mining engineers, the American Institute of Mining Engineers was concerned with almost every technical facet of mine safety, and its relationship to the bureau was correspondingly more broad, complex, and, ultimately, more disappointing than was that of the NFPA. Cooperation began in 1913, when Holmes, in response to the request of AIME's president, Charles F. Rand, provided the institute with a list of possible members for a proposed Institute Committee on Coal and Coke. Soon the mine-safety functions of this committee were transferred to special Advisory Committees on Mine Explosions, Mine Supports, Mine Ventilation, Mine Fires, and Explosives, which were staffed by the AIME at the request of the bureau sometime in 1914. Optimistic about the setup but concerned about the small number of coal-mining men on important AIME committees, Holmes pressed the knowledgeable H. M. Chance into service on the AIME Committee on Mine Explosions, an area in which he was anxious to get the benefit of disinterested outside criticism.[92] The chairman of each advisory committee was also the chairman of a subcommittee to facilitate cooperation with the bureau. Holmes anticipated that the advisory committees would submit reports on their work to the bureau for possible publication. Nothing came of these plans. Spread throughout the country, committee members could seldom meet personally. When Manning took over for Holmes in 1915 he wrote to William L. Saunders, president of the AIME: "I am frank to say that the Bureau of Mines has not, since these committees were appointed a year ago, called upon them for advice and assistance; but this is a condition which, with your help, I propose to remedy in the future."[93] Manning was no more successful than Holmes, and in March 1916 Rice sent Manning a memorandum which opened with this evaluation of the AIME advisory committees: "As far as I know these committees have never taken any action." Rice felt that the situation could be improved if each committee were to have as secretary someone who was a member of both the AIME and the bureau, but there is no indication that Rice had any success in making these committees workable by the end of the decade.[94]

The attempt at cooperation with the American Institute of Electrical Engineers (AIEE) on the problem of electricity in mining was equally unproductive. Holmes had first suggested a three-way cooperative mechanism (bureau, AIEE, and AIME) for electricity in mining in early 1914. Everyone seemed to agree on the concept, but a year later no cooperation had been initiated and there had been no joint meetings. Three years after the original attempts to cooperate, Manning was still looking for such a relationship and was not even aware that the AIEE had a Committee on Electricity in Mining, even though its chairman was a bureau employee.[95]

Frustrating in a different way were the bureau's cooperative relationships with organizations in the testing field. Here the bureau had traditionally done its own work, as in the testing of explosives and motors. The American Society for Testing Materials maintained a committee (including several bureau officials) which advised the bureau on testing procedures. In 1914, however, the National Council for Industrial Safety, a major safety organization with a mine-safety section, said products, to meet its standards, would be sent for testing to either the bureau or Underwriters' Laboratories, a private testing organization. Holmes considered this to be unnecessary duplication and "likely to develop conflicting results and other unnecessary complications."[96] Within six months the problem had been solved to the bureau's satisfaction: a few products, most of them in the line of fire prevention, would be tested by Underwriters' Laboratories, which would include in its permissible lists all those devices which the bureau had tested; all other products would be tested by the bureau, which would publish a comprehensive list of all approved mining products, no matter where the testing took place and which would, in effect, pass on all tests made anywhere in the country on mine products. Difficulties were overcome with relative ease in this cooperative area.[97]

Unique for its cooperative relationships was the mine-rescue and first-aid field, where federal, state, and private agencies joined with the Bureau of Mines to develop what by the end of the decade was a good system. The Illinois Mine Rescue Station Commission, established in 1910 following the Cherry disaster, consisted of representatives from the bureau and the Department of Mining of the University of Illinois, as well as state mine inspectors, coal miners, and mine operators. When, in 1915, a statewide Illinois Mine Safety Association was created through the work of W. D. Ryan, the bureau was represented along with operators, miners, and the rescue commission.[98] An effective rescue and first-aid program also required cooperative relationships with private organizations and federal agencies. The federal Public Health Service provided medical and surgical employees for the mine-rescue cars operated by the bureau, and through the Interstate Commerce Commission the cars had free access to the rails. A private organization, the American Red Cross, in addition to its regular sponsorship of first-aid and rescue meets, in 1913 created a subcommittee on mines as part of its National First Aid Committee. Under the chairmanship of Van H. Manning, the subcommittee served as liaison to the bureau. The American Medical Association (AMA), the first organization to suggest that each bureau rescue car carry a surgeon, was particularly important for its work on the problem of resuscitation of persons overcome by mine gases. The bureau and the AMA made separate inquiries into this subject and in 1912 a cooperative Committee on Resuscitation was formed at the urging of Holmes, who valued the opinions and counsel of the AMA and cultivated a close relationship with the organization.[99]

The mine-rescue and first-aid arena was also the fertile ground in which the first national organization devoted solely to mine safety—the American Mine Safety Association (AMSA)—took root. The AMSA grew out of the first National Mine Rescue and First Aid Conference, called by Holmes and attended, in September 1912, by operators, miners, surgeons, inspectors, and heads of safety and inspection departments. According to its constitution and in practice, the AMSA was aimed almost entirely at the mine-rescue and first-aid field, its standing committees including Rescue Apparatus, Rescue Operations, First Aid Methods, First Aid Training and Contests, and Hospitals and Training Stations. Perhaps the most time-consuming of all its activities was the sponsorship of first-aid and rescue contests in all parts of the nation, usually in conjunction with organizations like the Kentucky State Mining Institute, the YMCA, the National Conservation Exposition, the American Red Cross, and the Appalachian Coal Operators' Association. Membership in the organization was open to individuals and organizations of all kinds, but coal operators dominated its proceedings.[100]

In spite of its relatively short, three-year duration, the AMSA's relationship to the bureau was particularly close. This was partly due to Holmes's interest in rescue work and first aid, but more importantly to H. M. Wilson, concurrently the first chairman of the AMSA and the engineer in charge of the bureau's Experiment Station in Pittsburgh. Wilson's dual position made cooperation with the bureau particularly easy and productive and was the beginning of his emergence as a figure of some stature in the mine-safety movement.[101]

While the bureau and the United Mine Workers coexisted comfortably, efforts to institutionalize an informal consulting relationship were not very fruitful. One attempt took place at Holmes's initiative in 1912; another grew out of a meeting of federal officials and union leaders in January 1914, called by Franklin Lane, interior secretary, and designed to provide discussion of ways to improve safety conditions in coal mines and of "what more active part the miners can be induced to take in this movement." This effort, too, was apparently unsatisfactory, for in 1917 another advisory committee was created, this time at the urging of W. D. Ryan, who had managed to convince the Executive Board of the UMWA of the need for a new commitment to cooperation.[102]

"I have fought persistently, and I trust successfully," Holmes wrote in late 1914 to one of the bureau's most persistent critics, "to keep the work of the Bureau of Mines entirely free from the influence of politics and all kinds of organizations."[103] Exactly what Holmes meant by this remark is not clear. Certainly he and Manning succeeded in preventing the bureau from becoming so closely tied to labor or management as to lose its credibility; the bureau had lost no powers to a bureau of labor safety; and it had continued to function as an objective scientific agency. In another sense, however, the essence of the Bureau of Mines in its first decade was political. Holmes and Manning went naturally to the political well to obtain for their bureau a broader grant of authority and funds for new and old bureau projects. In this process, the cause of coal-mining safety was promoted only indirectly, as the strength of the bureau grew; the new act of 1913 was both a reflection and a foreshadowing of the new energies and directions of the bureau in metal-mining safety, precious metal investigations, and conservation, and most of the funds appropriated under the 1915 Kern-Foster legislation went to the West and the metal mines, not to the coal mines. Bureau chiefs took the possibility of a bureau of labor safety as a serious political challenge, which it was. And in all these struggles, the bureau depended on the political support and influence of its friends on the outside—the United Mine Workers, the Association of State Geologists, the American Mining Congress, and the influential coal-mine operators. At times this support was purchased in the most elemental ways—by promising specific benefits in return for votes.

The ideal bureau, according to Hunter Dupree, although it might begin as a research institution, soon branches into two directions, on the one hand furnishing routine services related to its problem and, on the other, becoming involved in regulation. In its first decade the bureau fulfilled only the first of these ideal characteristics, obtaining regulatory powers only fleetingly during the World War and then in an area basically divorced from coal-mining safety or even mining safety. The lack of regulatory powers turned the bureau into an educational institution dependent on its ability to influence operators, miners, and others who could take action to promote safety.[104]

Another consequence of this powerless condition was the development by the bureau of a series of cooperative relationships, formal and informal, with state and federal governmental bodies and private organizations. With some exceptions these relationships were fruitful, not only in extending the political base of the bureau but in spreading the gospel of mine safety. These cooperative relationships also exemplify a broad principle governing the bureau's work in coal-mining safety during these years: the emphasis, almost without exception, was on process rather than content, on communication of knowledge rather than knowledge itself, and on communicating solutions to problems rather than problem solving in a scientific way. The bureau's research was primary only in the sense that it had to come first; it was secondary in the sense that all the bureau's scientific investigations were meaningless unless miners, operators, inspectors, engineers, foremen, and the states acted upon them. Thus this period of ferment in education was also one in which the bureau was intensely conscious of the importance of effective communication and the methods by which it could be facilitated. This emphasis had its parallel in the bureaucracy, for Manning was not a scientist and Holmes was not a great one. Both were administrators and facilitators, and Holmes, particularly, was nothing less than a bureaucratic entrepreneur.

CHAPTER

3

Legislation and Enforcement in the States

COAL-MINING SAFETY seemed to fit neatly into the American constitutional system. State courts and the United States Supreme Court had recognized early that health and safety were proper functions of the state police power, and most of the coal-producing states enacted some kind of coal-mining safety legislation in the 1870s. Major revisions took place in the 1880s in Illinois, Pennsylvania, Ohio, and West Virginia. Legislative activity slowed considerably in the 1890s, perhaps because of the depression; of the major coal-mining states, only Pennsylvania undertook a major revision of its laws in that decade. As of 1900, Pennsylvania and Illinois had detailed coal-mining statutes, generally recognized as the best in the nation. West Virginia's code was vague and brief, the worst of the lot, while Ohio's statutes occupied a middle ground. Before 1900, when explosions and disasters had threatened this comfortable legislative edifice, the states had responded with additional legislation and new enforcement measures. As mine-safety problems worsened during the first decade of the twentieth century, the basic solution appeared to remain the same: the states continued to shoulder the burden, considerably modernizing their legislation from 1905 through 1915. An emphasis on continuity, however, masks important changes in approach. In several states, the political decision-making structure was modified. Whereas nineteenth-century coal-mining safety laws had been the product of an operator-miner struggle within the legislature, twentieth-century legislation was usually developed by a commission of operators, miners, and other officials. Only when the interest groups agreed was the legislation transmitted to the regular political channels. More important, the Progressive years were witness to an attempt to nationalize the federal system, first through the Bureau of Mines, which was designed, in part, to bring rationality and uniformity to state legislation, and then through a related campaign for uniform state coal-mining safety legislation. Ultimately this effort to upgrade state safety legislation without essentially modifying the federal system proved a dismal failure, and had it succeeded, there is some doubt that the expected additions to state safety codes would have had much impact on mining safety. The obstacles to full enforcement were substantial.

As of 1900, the peculiar West Virginia combination of powerful operators and virtually nonexistent unions had left the state with a rudimentary mine-safety law, consisting largely of provisions too basic to be omitted. Operators were required to provide maps of their mines, but no provision was made for securing maps if operators failed to furnish them voluntarily. They were to furnish a minimum amount of air per minute per person working in the mine, but the law contained few instructions on how to get that air efficiently to the miners below. Operators had to employ fire bosses and water coal dust only if a mine generated gas in dangerous quantities. West Virginia law required each operator to employ a mining boss (foreman), but his legal responsibility was not clear and his duties were defined imprecisely. He was, for example, required to "examine every working place in the mine as often as practicable," a provision so vague as to allow him to establish his own priorities, even if they were inconsistent with safety.[1] "West Virginia," wrote the *United Mine Workers Journal* following the March 1900 Red Ash explosion, "is noted in the mining world for her insufficiency of proper laws for the insurance and protection of her miners."[2]

This evaluation had little impact among West Virginia lawmakers, who consistently frustrated efforts to improve the law. Led by Governor G. W. Atkinson ("it is but the natural course of mining events that men should be injured and killed by accidents"),[3] state senators and representatives defeated a bill which would have made the mine boss the agent of the operator, thereby shifting legal responsibility; another bill which would have provided for examination of state inspectors, mine managers, mine examiners, and hoisting engineers; a third bill which would have made neglectful operators guilty of a felony and subject to large fines; and a fourth calling for a license tax on the mining of coal, partial proceeds to be used for inspection services and miners' hospitals.[4] This negativism is understandable in light of a general hostility to legislation among coal operators and, more surprisingly, West Virginia inspectors. In 1900 West Virginia operators were surveyed with regard to their views on legislation. Of 118 operators reporting, only thirteen recommended more legislation, and fifty-two were "adverse to legislation." Of the thirteen favorable respondents, not one suggested changes in ventilation requirements. Operators favored state examination of mine bosses and fire bosses by a two-to-one margin, but for some reason this did not come within their definition of "legislation."[5]

Led by chief inspector James Paul, West Virginia inspectors even opposed technical examinations for mine and fire bosses. Paul suggested that such examinations would make it impossible for miners to fill the positions; District 1 inspector I. M. Kelley raised the expert bugaboo, charging that under a similar Pennsylvania law, bosses "are more competent to explain why a certain accident happens than to foresee and prevent the accident."[6] Although Kelley, Paul, and other inspectors made occasional suggestions for new legislation in their reports, until about 1905 they were generally satisfied with the West Virginia legal code and convinced that the major mine-safety problem lay with the miners. Referring to the reluctance of miners to get out from under dangerous roofs, Paul noted: "Such accidents are little short of deliberate suicide. No legislation can reach such cases as this."[7]

By 1905 sentiment was building toward a major revision of the law. Responding to a series of relatively small mine disasters in West Virginia and the 1906 Courrières explosion, Paul dropped his opposition to legislation and took an activist role. To counter the problem of miner inexperience, he was now ready to request legislation providing for state examination of mine and fire bosses. In addition, he called for employment of shot-firers and more inspectors and requested that inspectors be given authority to close a mine operating outside the law and to require use of safety lamps in dangerous mines. Recognizing the difficulty of developing legislation satisfactory to both operators and miners "since the parties concerned desire absolvence from any responsibility," Paul asked the governor to appoint a commission of expert mine men—operators and miners—to review conditions and draft a bill for presentation to the legislature.[8] The governor took his suggestion sometime in 1906, and the result was the passage in February 1907 of a substantial revision of West Virginia's mine-safety laws, unfortunately weakened by the elimination of several amendments. The new law brought some precision to the duties of mine management. The mine foreman or his assistant was now charged with instructing inexperienced miners until they were familiar with dangers, and with measuring air currents twice each month at inlet and outlet. In another provision designed to deal with the problems of inexperience and carelessness, the mine foreman or assistant was required to visit and examine every working place in the mine at least every other day. The act contained a weak statement requiring watering and prevention of dust accumulations "as far as practicable," required that the mine should be worked exclusively by locked safety lamps when explosive gas was present in dangerous quantities, and increased ventilation requirements. Although the district inspectors could not close mines on their own authority, they could now appeal to the chief inspector to do so, and they could now also demand that more air be pumped into a mine.[9] Interestingly, the idea of examination for mine employees, popular with operators and advocated by the chief inspector, was in the Mining Commission Report but was not included in the final legislation. A provision authorizing the district mine inspector to "prescribe the condition under which such solid shooting may be done" was an improvement in the old law but considerably weaker than the bill's original provision virtually prohibiting shooting off the solid. Amendments to restrict shot-firing were also soundly defeated.[10]

Less than a month before this legislation was passed, an explosion at Stuart, West Virginia, killed eighty-four men, and another at Thomas, West Virginia, killed twenty-five. On February 4, 1907, two days after the Thomas explosion, the House and Senate passed a Joint Resolution providing for a legislative committee to investigate both the Stuart mine disaster and the state Bureau of Mine Inspection and to determine what measures might be taken to prevent future disasters and to insure more stringent enforcement of the mining laws. In a message requesting the investigation, Governor William M. O. Dawson cited certain illegal characteristics of the Stuart operation in Fayette County and stressed the possibility that the Bureau of Mine Inspection might have been "inefficient or derelict in its duty."[11] The committee spent well over a year investigating mine safety in West Virginia. It made six trips to different parts of the state, securing testimony on the Thomas, Whipples, Stuart, and Monongah mine explosions. The absence of a completed report placed Dawson in a difficult position in early 1908. Monongah and other disasters, said the governor in his annual message, had given this state "a bad eminence in this regard." Without the committee's report he felt "some embarrassment in referring to additional legislation" but suggested that shooting off the solid might be prohibited and mine officials certified. To complete the picture of confusion, he offered the hackneyed suggestion that enforcement of existing law, rather than new law, was the real need. Although he had not yet received the report of the joint investigative committee, Dawson was already setting up a barrier against new legislation. "Any unreasonable burden placed upon our coal industry," he said, "would result in shutting down our mines or a part of them, or reduce the wages of miners and other workmen."[12] In the meantime, Paul was pursuing his new activist posture. To the extra session of the 1908 legislature he presented a list of sixty-three proposed amendments to the mining law, including a number which would have increased the enforcement powers of the inspectors. Paul's district inspectors were also becoming increasingly aggressive in their demands for additional legislation.

With the publication of the report of the investigative committee, this series of events rapidly came to a conclusion. The committee had appropriately focused on the explosion problem and the coal-dust phenomenon. It had arrived at the reasonable conclusion that a major cause of explosions was the ignition of coal dust by blown-out shots, the latter the result of excessive amounts of solid-shooting by miners. At this point, however, the committee lost touch with the issue. Instead of suggesting any number of possible remedies, as Pennsylvania and Ohio inspectors had done after similar investigations at Monongah, the committee decided that new legislation was no answer. "These evils," read the final report, "cannot be corrected by statutory law but must be accomplished by and through the vigilance of those who are in charge of the work, aided by that earnest and practical training which will fully impress upon the mind of each and every individual working therein the danger to his own life as well as the lives of those associated with him."[13] One member of the committee asked to have his name withdrawn from the final report. A convention of Subdistrict 1 of District 17 of the United Mine Workers, at Charleston, called the report a whitewash and condemned the committee for not allowing a minority version.[14] The result, however, was predictable. Buoyed by operators who wanted no additional legislation, the West Virginia legislature rejected every mine-safety law proposed in 1908 and passed no major piece of safety legislation for the next seven years. In contrast, it is worth noting that the West Virginia House and Senate overwhelmingly approved resolutions calling for a national bureau of mines.[15]

The 1915 legislation brought some substantial improvements, many of them long overdue. Six years after the Cherry mine fire had raised the issue in sharp relief, West Virginia had its first provisions for preventing mine fires in underground stables; the articles dealing with electricity and machinery were similarly late. Emphasis on shot-firing had produced a new requirement that no shots could be fired in any working place known to liberate explosive gas until some competent person had examined the area, but the law still contained no adequate restrictions on solid-shooting and the miners, rather than professionals, were still doing the shooting. Ten years after the chief inspector had suggested such a reform, foremen and fire bosses were required to pass state examinations and receive state certification.[16] Perhaps more interesting than the provisions of the 1915 legislation are developments that occurred in the administrative structure of mine safety in the years between the passage of the two major acts. Frustrated by the absence of legislation, chief inspector John Laing began to issue administrative regulations having, in his opinion, the force of law. Laing chose this method to deal with the problem of coal dust, insisting upon use of shot-firers and permissible explosives in unwatered dusty mines. In an even bolder move, Laing went beyond statutory requirements and administered departmental examinations for foremen and fire bosses. Denied the sanction of legislation, the West Virginia Department of Mines had opened up an important new area of regulation.[17]

In 1900 Ohio had the vaguest safety legislation of the major coal-mining states. Nearly all its major provisions dated to the 1880s. District inspectors were required to examine all mines "as often as possible"; ventilation requirements were minimal. The state's regulations with regard to machinery, hoisting, and haulage were uniquely lacking in precision and detail. The law did, however, contain several redemptive features. It obliged the inspector to initiate prosecution when operators failed to correct deficiencies, and it gave the inspector leeway to go beyond the statutes and define additional dangerous conditions for himself. Where West Virginia law had given the coroner the option of holding an inquest into the causes of mine deaths, Ohio required the coroner to do so. Finally, Ohio law, in contrast to West Virginia's, contained several provisions which were protective of the rights of mine labor. One section maintained the right of legal action for those injured through violation of the safety laws or for relatives of those who died; another contained elaborate procedures for removing incompetent inspectors from office; and a third, the most unusual, sanctioned the right of employed miners to appoint two of their number to inspect the mine once each month.[18]

For some time, Ohio legislators showed no interest in bringing the state's safety legislation up to reasonable standards. A bill to improve mine ventilation was passed only by the state House of Representatives; another which would have required shot-firers and fined operators who refused to furnish them received approval only in the Senate. Bills for the taxation of coal operations, for the regulation of explosives and blasting, and for upgrading legislation on mine machinery and appliances, never emerged from committee. George Harrison, the state's articulate and active chief inspector, recommended revisions in 1905, following his first year in office, but the legislature ignored his pleas that the state put something on the books to deal with electricity. "The Mining Laws of Ohio," noted Harrison in a 1906 letter to district mine inspectors, "are so mild and far from covering the conditions of mining at the present time, that it is certainly no hardship for operators to comply with them."[19]

Although Ohio was completely unlike West Virginia in that the state had not suffered a major disaster (five or more dead), revival of interest in safety legislation came roughly at the same time in both states. In 1906 Harrison obtained the assistance of Governor Myron T. Herrick and, in spite of determined opposition from operators before the Senate Committee on Mines and Mining, secured two pieces of legislation of some importance, given existing Ohio law. One called for watering and removal of coal dust, charged the inspector with insuring that dust did not collect, and provided fairly heavy fines for operator failure to comply. The second, a miner competency act, required a prospective miner either to produce evidence that he had worked one year as, or with, a practical coal miner, or to be accompanied in his work by a competent coal miner.[20] Disturbed by the failure of the legislature to act on other basic matters such as electricity, explosives, and solid-shooting, in April 1907 Harrison, like his counterpart in West Virginia, turned to administrative law and issued a series of safety regulations. "This notice," said Harrison, "was issued in the utter absence and lack of power authorized by the present mining laws which have long been in existence, and which utterly fail to cover many of the dangers." The regulations covered blasting and electricity and required operators to employ fire bosses in all mines generating explosive gas. A year later the electricity and explosives regulations were codified and a requirement that each mine be inspected every three months was added to state laws.[21]

The explosions of 1907, particularly at the Monongah mines in West Virginia and the Darr mines at Jacobs Creek, Pennsylvania, imparted a new momentum to the Ohio safety movement. Implementing a suggestion made by Harrison almost four years earlier, in March 1908 the legislature, with the strong support of state operators, created a commission consisting of three practical miners, three operators, and one other person (ultimately Harrison), to investigate safety conditions and make recommendations for changes in the mine law. The commission visited over thirty mines and met with Joseph A. Holmes of the Geological Survey and the team of European experts which Holmes had imported. Because of the dated nature of so much of Ohio safety law, the commission decided to draft an entirely new code which, if enacted, would repeal all existing legislation. After eighteen months of deliberation, the commission submitted its report to Governor Judson Harmon. In a special message to the legislature which clearly revealed the influence of the recent mine fire at Cherry, Illinois, Harmon transmitted the commission's recommendations to a receptive legislature. They became law without a change and without a single dissenting voice or vote in House or Senate.[22]

For the time, the new Ohio legislation was strong. New ventilation requirements brought the state to parity with Illinois and Pennsylvania. Although operators could still use the inherently dangerous furnace for ventilation, they could no longer employ air shafts with bottom furnaces as a means of ingress or egress. Another section reflecting the Cherry mine fire regulated underground stables and livestock, and the act contained precise provisions for signaling, haulage, and electricity. In one important area, the commission differentiated Ohio law from that of other states—Ohio's law was the best of the major coal-mining states in providing supervision and training for inexperienced miners. Not only was the foreman to visit each working place at least once every other day (with authority to order work stoppage), but the law also called for an "overseer," a kind of foreman for the inexperienced. Until the inexperienced operative had been in the mine three months, the overseer was required to visit his working place at least once every four hours; from three to six months, once every six hours; and from six to nine months, at least once a day. The new legislation also attempted legislative supervision of miners by providing instructions for mining and shooting the coal.[23]

Some weaknesses in the new Ohio law were evident at once. A district inspector pointed out that the new mining code did not deal adequately with the increasingly serious problem of shooting off the solid. Another, while admitting the progressive nature of much of the Ohio legislation, suggested that in several areas, including the certification of foremen and fire bosses, "it has been content to remain passive."[24]

A governor-appointed Ohio Coal Mining Commission in 1913 added its concern that a change to the mine-run system of payment, then under consideration, would increase fatalities from falls of roof and coal. Though the commission advocated mine-run for reasons of conservation, it balanced this suggestion with a number of safety recommendations. It called for legislation requiring that every company employing more than thirty-five men at the working face employ a safety foreman to fire the shots or supervise their firing. In its strongest statement, it recommended that solid-shooting be forbidden by law except in those mines where it was absolutely essential. The commission rejected examinations for miners but favored them for those in positions of responsibility. Finally, the commission recommended that coal-mining safety functions be placed under a regulatory commission with discretionary authority.[25]

These proposals were welcomed by the miners of the state but met an unfavorable reception at the hands of operators. High costs, always at issue when legislation calling for new personnel was involved, were a major factor in operator opposition. *Colliery Engineer*, on the other hand, focused its attack on the proposed industrial commission, insisting that local conditions should determine mining regulations: "To substitute the opinion of a commission at Columbus for the facts shown by local conditions prevailing at the mine, is the height of absurdity. The idea is a socialistic one evidently put forth for political purposes." This journal editorial argued that the employment of safety foremen would destroy discipline and opposed the pending solid-shooting legislation because it entailed Industrial Commission discretion.[26] Operator objections proved largely futile. In 1914 mine safety was placed under the Industrial Commission of Ohio and specifically under the conscientious new safety commissioner of mines, J. M. Roan. Roan immediately prepared a list of mine-safety rules. The legislature also prohibited solid-shooting without a permit from the Industrial Commission, the law providing that permits could be issued only if solid-shooting were necessary for the "just and reasonably profitable operation of such mine," and as of February 1914, Ohio miners were being paid by the mine-run method. Operators succeeded only in preventing passage of safety-foremen legislation.[27]

With the possible exception of Pennsylvania, Illinois had the best mine-safety laws in the nation.[28] By 1899 the state had provided for examination and certification of all mine officials, and in 1908 Illinois became the first state to extend certification to miners. Illinois's legislation carefully defined the duties of mine personnel, though its instructions for examination of the mine by the fire boss were the most backward of any major coal-mining state. Illinois was perhaps the only state to approach the problem of miner carelessness and inexperience through the avenue of formal education, providing in 1911 for miners' and mechanics' institutes to instruct miners in the intricacies of gas, dust, and other safety matters. This promising innovation was lost in 1915 when funds appropriated for the project were vetoed. The state's mine-fire regulations were also unique, a product of the 1909 Cherry fire.

Illinois was unusual in that an inordinately large number of the legislative struggles over mine safety occurred over issues related to shot-firing, an outgrowth of the method employed in measuring miners' wages. After 1897, Illinois miners were paid on a mine-run basis; that is, they were paid for all the coal brought out of the mine, rather than for just those chunks that would not pass through a wire screen. When miners no longer had to preserve the coal in chunks, they stopped undermining the coal before shooting it and began to use more powder. Force replaced finesse at the working face. Operators and miners were soon poised on opposite sides of the question, the former favoring abolition of the mine-run system and legislation requiring miners to undercut the coal, the latter insisting that operators employ shot-firers to work when men were out of the mine. The first in a series of decisions went to the miners in 1905, when the legislature, responding to the Ziegler mine disaster of April 3 and acting at the close of a legislative session when operator representatives had left, passed a shot-firers bill. David Ross, the equivalent of a chief inspector in Illinois, was convinced that it was not enough to shift the firing function to professionals; he recommended divorcing miners completely from blasting and handling of powder. Although this suggestion, which would have turned the average mine worker into a coal loader, was not acted upon, the legislature hardly dropped the matter. The basic shot-firing law was amended often and was complemented by several important pieces of legislation regulating the handling and the quality of explosives.[29]

The distinctive characteristics of Illinois law should not obscure the essential similarity to the reform process in other states. Reform took place within the same basic time frame, roughly 1905 to 1911, when Illinois's major omnibus bill became law. And here, as elsewhere, operators, represented in the Illinois Coal Operators' Association, and miners, through the United Mine Workers (rather than the Illinois State Federation of Labor), played predictable roles from which they seldom strayed. Operators opposed safety legislation which would increase mining costs or restrict their control over the labor force; miners favored most safety legislation, choosing to oppose it only when it affected their paychecks. Such was the case in 1908, when the miners interpreted a mine-inspector proposal which would have prohibited solid-shooting under certain conditions as an attempt to increase work load without increasing worker compensation.[30] Although political compromise between the two contending groups could and did occur in all periods, until a formal mechanism for compromise was found, new legislation often left bitterness in its wake. That formal mechanism was the legislative commission, of which Illinois had two. The Joint Powder Commission contained representatives of operators and miners and functioned as a fact-finding board for dealing with blasting powders, an issue of particular complexity and sensitivity in Illinois. The Mining Investigation Commission, created under law in 1909 largely at the urging of operators, was intended as an ultimate board of compromise for all coal-mining laws, including those dealing with powder. Its membership of three operators, three miners, and three nonpolitical independents was remarkably successful in balancing the disparate interests of operators and miners, profits and safety. Only once, in 1914, did the miners appear seriously interested in abrogating the commission arrangement and taking their case directly to the legislature.[31]

In 1900 Pennsylvania mines were operating under 1893 safety legislation which, though exceedingly advanced for its time and pioneering in a number of areas, was rapidly becoming obsolete under turn-of-the-century mining conditions. Coal-trade journals had mixed opinions. Some were skeptical of the ability of new legislation to handle the problems ("The question may be asked—What's the use? We have had various laws.");[32] others, indeed the majority, supported revisions in the mine-safety law.[33] Led by its forceful and diligent chief, James Roderick, the Pennsylvania inspection service soon made its presence felt in the reform camp. Roderick regularly pointed out glaring deficiencies in the safety law. Years before West Virginia's inspection department became concerned about deaths from falls of roof and coal, he was advocating systematic regulations for roof propping. Among his early causes were reforms in the laws governing electricity, foreman supervision of miners, ventilation, responsibilities of superintendents, and solid-shooting.[34] The chief inspector's activism carried into administrative regulation, where, as in West Virginia and Ohio, inadequacies in the written law were met by administrative discretion. In 1907 discretion took the form of a list of department rules going beyond the statutes, but Roderick's greatest administrative triumph involved interpretation of existing law. The case began with the concern of Isaac G. Roby, inspector in the Connellsville coke region, where solid-shooting had become a regular practice because it was peculiarly suited to mining coking coal. Roby asked Roderick to interpret the 1893 legislation, which required that coal be "properly" undermined. Roderick stretched the provision to its limits, holding that proper undermining meant complete undermining; no shots could be drilled deeper than the undercuts. Neither operators nor miners liked the decision, and it was largely ignored. Rather than acquiesce, Roderick clung to his interpretation of the law and, within two years, had convinced the Connellsville operators voluntarily to employ shot-firers to do the shooting when the men were out of the mine. "In the opinion of the Department," wrote Roderick, "no other single movement has been inaugurated within the history of the coking business that gives promise of such fruitful results in securing safety to the employes."[35]

The sequence of events that was to culminate in 1911 in a major revision of Pennsylvania law began in 1907 when the state legislature, acting on one of Roderick's three-year-old suggestions, called on Governor Edwin S. Stuart to appoint a five-man commission (two operators, two miners, and Roderick) to study the state's mining laws. Presented to the legislature in early 1909, the proposed code encountered opposition from two sources. As expected, operators emphasized the extent to which the new law would handicap the state in competitive markets by increasing costs of production. While the commissioners conceded that if adopted their report would "be a hardship to the operators' interests" and increase operating costs by one to five cents per ton, operators placed the cost increases at six to sixteen cents per ton. Appearing at hearings over the code, a representative of the Coke Producers Association of the Connellsville Coke Region said interstate competition would make it impossible to pass increased costs on to the consumer and that costs would instead be absorbed through wage cuts. Having reached this point, he concluded his statement with an appeal from William Jennings Bryan—"you shall not press down upon the brow of labor this crown of thorns"—that must have seemed quite out of character to the committee and bystanders.[36] Operators also objected to provisions in the law which they felt would mean competency examinations for miners, and to others which placed too much authority in the hands of the mine foreman. "The only thing left for the operator," said one representative of the Butler-Mercer field, "is to dig up enough money for the payroll and pay for the damages done under the supervision of the licensed men furnished by the state."[37]

Although the concerted opposition of operators at the House hearings was probably enough to kill the bill, its fate was sealed when Francis Feehan, one of the commission's labor representatives, refused to sign the final commission report because of two articles which he thought insufficient. Feehan wanted broader restrictions on the use of open lights and complete prohibition of the use of electricity in gaseous mines. Roderick, who had in other circumstances advocated this view of electricity, now saw Feehan and his supporters as a "radical" element which failed to appreciate the necessity for compromise. To the credit of the chief inspector, there is evidence that Feehan's stand was not representative of miner opinion. Patrick Gilday, authorized to speak for 37,000 miners in Pennsylvania's District 2, faulted the bill for its "commercialism" and for its general inadequacy, but he pressed for its release from committee.[38]

Following new discussions with operators and miners, much the same bill which in 1909 failed to clear the House Committee on Mines and Mining was introduced into the next legislature in 1911 and passed with little discussion. For some reason, perhaps new legislation in Ohio and Illinois, Pennsylvania operators did not oppose the bill. *Mines and Minerals* speculated that there must have been some mistake or miscalculation: "a careful and critical analysis of the 1911 Bituminous Mine Law of Pennsylvania cannot do otherwise than awaken a suspicion that the bituminous operators who sanctioned its passage did not fully realize to what extent the law would increase their cost of production."[39] The 1911 law gave Pennsylvania the most detailed, and in many ways the best, mine-safety legislation in the nation. The articles on electricity, safety catches, hoisting machinery, openings, and motors were models of precision, designed, it would seem, to codify what might normally be administrative regulations. The law reflected Roderick's ongoing concern over explosions and his knowledge of European and American experiments on the coal-dust phenomenon. In spite of his earlier hostility to a national bureau of mines on the grounds that "the persons in charge of the Bureau would unquestionably know less about the real condition of the mines than do the managers, superintendents, foremen and inspectors,"[40] Roderick was sympathetic to the bureau's work in dust and explosives. As a result, the 1911 law contained some of the most far-reaching coal-dust regulations yet enacted. It required that no shot-firer could fire a shot unless the dust was thoroughly wet for a distance of eighty feet around the hole; it prohibited the use of arc lamps in the presence of coal dust; and, in a regulation unique for the major coal-mining states, it mandated the use of permissible explosives, as tested by the Bureau of Mines, in portions of dry and dusty mines where explosive gas was present. Under the same conditions of gas and dust, the foreman was required to exercise extra care in firing. Along with operators and inspectors, Roderick had insisted for years that the foreign mining population was a safety hazard. Although the 1911 act did not fulfill Roderick's wish to prohibit employment of persons without knowledge of English in dusty and gaseous mines, it contained several provisions designed to alleviate the foreign-language problem. One, for example, required that all danger signals be explained to non-English speaking employees, through an interpreter if necessary. Neither Ohio nor West Virginia codes contained foreign-language provisions.[41]

The act's most definitive characteristic was the active nature of its language, a fact of considerable importance in fixing legal responsibility and, therefore, in enforcing the act itself. Where the laws of Illinois, Ohio, and West Virginia had been passive (e.g., coal is to be properly mined), Pennsylvania law made mine employees active agents in the safety process (e.g., the foreman is to insure that coal is properly mined). Everyone who had any function within a mine—miners, drivers, trip-riders, cagers, and others—was given safety responsibilities. The superintendent, a kind of business manager for the mine and the direct representative of the owner, was charged with enforcing the mine law upon instructions from the inspectors and was also prohibited from obstructing the mine foreman or other officials in the performance of duties under the law. Such a provision was believed necessary because superintendents were traditionally interested not in safety but in production and profits. The mine foreman, however, shouldered the brunt of the safety responsibility. It was his task to supervise the working places in the mine, maintain safe haulage roads, make sure the coal was properly mined, and keep the dust watered; and he was responsible for hiring most of the officials below him, including the miners and the fire bosses. Although the miners, inspectors, and other reformers would have preferred that the superintendent shoulder more legal responsibility, the active language of the act was itself a major step forward in mine-safety legislation.[42]

The 1911 act remained Pennsylvania's basic mine-safety law through at least 1920. During the Progressive years, almost every attempt to amend it was defeated, the only important amendments coming in 1915 when provisions for certification of foremen and fire bosses were deleted from the law due to a new workmen's compensation act. Unfortunately, the basic enlightenment of the 1911 law was not matched in the state's enforcement procedures. An inspector could remove workers from an unsafe mine if the danger was extraordinary, but he could not otherwise close a mine except by going through a four-stage process culminating in an appeal for a court injunction.[43]

The law really exists only within the context of those who have some interest in its application and enforcement—operators, superintendents, foremen, fire bosses, mine workers, inspectors, judges, legislators. Each group interprets the law, and each contributes to its fulfillment or its decay. To what extent were these groups committed to mine safety? To what degree was state mine-safety legislation enforced?

Money is one measure of commitment. Presumably a state generous in its mine-safety appropriations is more committed to enforcing its statutes than its miserly neighbor, though the latter may have better legislation. Continuing increases in appropriations by several states would indicate a general commitment to enforcement. Table 4 presents the total yearly expenditures for the mine-inspection departments of Pennsylvania, West Virginia, and Ohio for a twenty-year period, expressed as a function of tonnage.[44] The figures fulfill some expectations and modify others. Of the three states, West Virginia consistently spent the least money per ton on mine safety. Though Ohio's safety legislation was inferior to Pennsylvania's, that state's legislators often voted more funds for enforcement than their counterparts to the east. The evenness of the figures before 1903–1904 (for Pennsylvania) and before 1906–1907 (for West Virginia and Ohio), indicates that mine safety was funded, but not taken very seriously, before those dates. Each of the states made a brief effort to redeem itself after the 1907 disasters, expenditures reaching temporary peaks in fiscal years 1908 and 1909. Taken together, the three states increased their real expenditures significantly after 1903.

Most district and chief inspectors agreed that however much money was being spent on inspection, the number of inspectors was inadequate. This complaint was particularly common in West Virginia, where chief inspector James Paul took every opportunity to plead for a larger inspection force. Although the legislature responded and tripled the force between 1904 and 1907, claims of inadequacy con-

TABLE 4

Expenditures of Mine Departments of Pennsylvania, West Virginia, and Ohio, 1900–1920 (expressed in 1926 dollars/1,000 short tons)

Fiscal year	Pa. bituminous	West Virginia	Ohio	Three-state average
1900	0.97	*	*	*
1901	0.94	0.85	1.30	0.99
1902	0.96	0.87	1.27	1.00
1903	0.92	0.72	1.28	0.94
1904	1.43	0.65	1.29	1.25
1905	1.19	0.70	1.24	1.10
1906	1.25	0.60	1.11	1.09
1907	1.16	1.07	0.92	1.11
1908	1.57	1.27	1.18	1.45
1909	1.33	0.98	1.43	1.26
1910	1.19	0.79	1.19	1.09
1911	1.24	0.93	1.72	1.23
1912	*	0.85	1.66	*
1913	*	0.87†	1.66	*
1914	1.55	0.89	3.47	1.51
1915	1.45	0.87	2.85	1.39
1916	1.28	0.72	*	*
1917	1.26	0.51	*	*
1918	0.69	0.44	*	*
1919	0.82	1.01	*	*
1920	*	0.69	*	*

*Not available
†An estimate, based on expenditures for nine months

tinued. In 1900 or 1915, West Virginia inspectors were not inspecting the mines as often as the law required.[45] The same was true of Kentucky, a state that rose to prominence as a coal producer in these years and one seemingly caught off guard by its own rapid growth. In 1892 Kentucky's eighty-eight operating mines, producing about 3 million short tons, were inspected by the chief and his assistant; in 1905, the same two officials were responsible for 203 mines with an output of well over 7 million short tons. Additions to the inspection force notwithstanding, Kentucky's chief inspector continued to insist on the physical impossibility of inspecting the state's mines as often as the law required with the personnel he had. The problem was especially acute in inspection districts with difficult topography and inefficient transportation.[46] Ohio's inspection service did not face these problems and managed to maintain a relatively high inspector/tonnage ratio, yet figures gathered there for 1898 and 1899 reveal how few of the state's mines were inspected four or more times in the course of a

TABLE 5

Frequency of Mine Inspection in Ohio, 1898 and 1899

Frequency of inspection	Number of mines 1898	1899
1	419	404
2	223	115
3	125	95
4	73	54
5	28	33
6	22	20
7	9	17
8	10	4
9	1	8
10	2	6
11		2
12		5
13		1
17	1	

year, a number recurring in the legislation of most states. Table 5 shows how often Ohio mines were inspected.[47] These figures also indicate that in the absence of a statutory requirement for regular inspection, the inspectors apparently went where they felt they were most needed—to the dustiest, most gaseous, most poorly ventilated mines. Presumably, the state's dozen most dangerous mines received more adequate inspection under this system than they did in other states, where each mine was, in terms of the number of times it had to be inspected, treated equally. In short, although the numbers reveal the need for more inspectors, Ohio's flexible use of manpower may have been the most efficient.

If we can believe Charles Connor, inspector for the Fifth District in Pennsylvania, that state, for all its fine legislation, did not transcend the inspection problem. Connor was supposed to visit each mine in his district once every seventy-eight working days, but the district contained 103 mines and it took Connor an average of one day to inspect each mine. What makes the Pennsylvania experience unique is that Connor's analysis, regularly reinforced by the state's miners and operators, did not produce significant additions to the inspection staff. Between 1900 and 1914, when other state inspection forces were growing rapidly, Pennsylvania's changed little. The state added one bituminous inspector to its force of twenty-four in 1911, another in 1912, and two more in 1913.[48] For this snaillike progress the state's chief inspector, James Roderick, must share responsibility. Roderick

TABLE 6
Tonnage per Inspector, 1900–1920
(in 1,000 short tons)

Year	Pennsylvania	Illinois	West Virginia	Ohio	Four-state average
1900	3,326	3,681	*	2,712	*
1901	3,429	3,904	6,017	2,992	3,682
1902	4,107	4,705	6,142	3,360	4,276
1903	4,296	5,279	7,334	3,548	4,624
1904	4,081	5,210	8,101	3,485	4,552
1905	4,933	3,843	7,558	3,650	4,786
1906	5,387	4,148	8,478	3,887	5,225
1907	6,255	5,132	9,618	4,591	6,123
1908	4,882	4,766	8,379	3,753	5,065
1909	5,748	5,090	10,369	2,794	5,482
1910	6,271	4,590	12,334	3,421	5,965
1911	5,782	4,473	11,966	2,563	5,348
1912	6,225	*	5,561	2,877	5,210†
1913	6,206	*	5,938	3,016	*
1914	5,285	*	5,976	1,570	*
1915	5,265	*	6,432	1,869	*
1916	5,676	*	5,764	2,894	*
1917	5,748	*	5,762	3,395	*
1918	5,951	*	5,995	3,817	*
1919	5,025	*	5,269	2,989	*
1920	5,686	*	4,735	3,823	*

*Not available
†Estimated

believed that coal operators were beginning to conceive of the inspector as a supervisor, as a substitute for hired, private supervision by the foreman, fire boss, and superintendent. He held firm to his belief that the state had enough inspectors, and that to increase their number would simply allow coal companies to rely on them more than the law contemplated. "It is not the duty of the inspector," Roderick said, "to oversee the work of the miners except at stated intervals. His important duty is to see that the persons in charge of the mines comply with the provisions of the law governing the operation of the mines so as to guarantee health and safety to those employed. The State cannot share the responsibility for the conduct of the employes. The responsibility must rest upon the superintendents and foremen and other officials of the mine."[49] If the number of inspectors is related to the amount of mining taking place (see Table 6),[50] it appears that Roderick may have overestimated the extent to which Pennsylvania's inspectors were taking the place of supervisory personnel. During the crucial years from 1905 through 1912, the work load of Pennsylvania inspectors was higher than that in Ohio or Illinois. West Virginia's inspectors were consistently responsible for far more tonnage than those of any other state, Ohio's for far less. Ohio's ratio benefited from the low growth rate of the state's bituminous industry; West Virginia's suffered from the opposite condition. As a group, the states failed to relieve the burdens on their inspectors until fiscal 1908.

The quality of state inspection staffs was seriously affected by low salaries. Impoverished assistant inspectors regularly resigned for the greener pastures of private enterprise, making it difficult for states to maintain experienced work forces. Chief inspector Harrison of Ohio estimated that it took two years to train a state inspector and lamented his state's salaries, which he said had not been raised in fourteen years. In Kentucky, C. J. Norwood lost his entire force of three assistants over a salary dispute in 1908. West Virginia experienced no mass defections, but resignations were common even among loyal employees like P. A. Grady, who in 1912 reluctantly tendered his resignation to accept a position with the Justus Collins coal enterprises. The state suffered in comparison with Pennsylvania, where inspector salaries were always higher, but West Virginia Governor John Cornwell came nearer the problem in his 1919 message. "It is an anomalous condition," he said, "that the State should require the Chief of the Department to secure and retain first class men as District Inspectors whose rate of pay is less than that of men who drive mules."[51]

Inspectors who resisted the lure of higher salaries were still unlikely to be totally objective in performing their duties. Upward mobility for a mine inspector meant either moving into the federal bureaucracy (as James Paul did in 1908) or, more likely, accepting a position as superintendent with a mining company at a salary between 50 and 100 percent above his wages as an inspector. Although a great majority of inspectors began their mining careers as miners, very few went directly from the pits to the inspection service. Most inspectors, even in Illinois where the force had a reputation for radicalism and strong ties to labor, came to their posts following employment as mine managers or superintendents. Chief inspectors were more likely than district inspectors to have some training in civil or mechanical engineering, perhaps a slight liberalizing influence. Engineering backgrounds were common to Robert Montgomery Haseltine, who served as Ohio's chief inspector from 1888 to 1898; James Paul, West Virginia's chief from 1897 through 1908; and the English-born Joseph C. Thompson, director of Illinois's Department of Mines and Minerals from 1918 to 1920. Certainly few chief inspectors were as unpopular with labor as West Virginia's John Laing, who began as a miner but at the time of his appointment owned several mining properties in the Cabin Creek and New River districts.[52]

Engineered by West Virginia operators, Laing's appointment is only the best example of the politics of mine inspection. That mine inspection was often political was an article of faith not only among miners but with operators and inspectors as well. If the *United Mine Workers Journal* overstated the case in its insistence that almost all mine inspection departments were "controlled absolutely by a ring of political ward heelers,"[53] inspection services were nonetheless strongly politicized, particularly in those states (West Virginia and Ohio but not Illinois or Pennsylvania) where appointments were made directly by the governor. In Ohio inspection posts were considered political plums for the party in power. As of 1911, chief inspector Harrison claimed the state had largely eliminated politics, but this was true only in the sense that the state's governors were now appointing an equal number of Republicans and Democrats. Party affiliation, in short, was still at issue in Ohio mine inspection. A West Virginia district inspector implied a very different definition of politics when he reported "that there are coal operators who will endeavor to have a district inspector removed from office rather than obey the mining laws, or carry out the recommendations made by the inspector."[54] It was this definition which regularly turned the appointment of mine inspectors into a struggle between rival interest groups, to the great disadvantage of the miners, who failed to exercise much influence even through their unions. To the operators went most of the spoils. "There is not an inspector in the state," said a frustrated Logan County, Illinois, miner, "who is not holding his job through the influence of some coal operator."[55] Testifying before the United States Industrial Commission, R. G. Brooks, a Scranton, Pennsylvania, coal operator, admitted that operator influence in the appointment process could prejudice mine inspection: "Oftentimes corporations use their influence for some person, and I think perhaps the average man will not be quite so severe where a corporation has aided him a little. I think that is another factor which is sometimes detrimental to the man's being absolutely free to carry out his inspections." Asked if the principle of competency was usually followed in selecting inspectors, Brooks replied: "No, I do not think it is."[56]

The mine workers viewed the politics of mine inspection as a function of the strength of organized labor in the coalfields. Inspection would improve when the workers were effectively organized to exert influence on the appointment process. But since organization promised no immediate relief, miners sought to limit operator influence in the appointment process by making the offices elective rather than appointive and, if that proved unrealistic, at least to insure qualified inspectors by requiring all candidates to pass board-administered examinations. Neither effort met with much success. Inspectors were elected in Kansas, Oklahoma, and the Pennsylvania anthracite region, but none of the major bituminous areas enacted the necessary legislation. Operators rationalized and argued that election would result in the selection of incompetent inspectors—and no one wanted that. Holmes opposed election on the grounds that it would politicize the system, a position which failed to recognize that inspection was already political.[57] Examination and certification of inspectors, a much less radical solution since it retained the appointment procedure, was used in Pennsylvania and Illinois but not in Ohio or West Virginia. In the latter state, bills for examination of mine inspectors were reported from committee with negative recommendations in 1901 and 1905, and an effort to create a state mining board in 1909 was strongly opposed by inspectors, who believed its composition—two operators and two engineers—would allow the operators to control the department.[58] While it is apparent that competency examinations did not eliminate politics, it is likely that the inspection systems of Ohio and West Virginia suffered from the absence of systems for eliminating candidates lacking appropriate technical abilities.

These procedures produced inspection systems stronger at the top than at the bottom. As a group, chief inspectors, no matter what their origins, did their best to enforce the laws. Only in West Virginia, where the highly political nature of John Laing's appointment as head of the state's inspection service in 1908 served to prejudice the agency in the eyes of miners, was the chief inspector criticized for insufficient devotion to law enforcement. Laing's administration, unfortunately, was unimaginative, even stagnant, and that of his successor, Earl Henry, was no better. According to one district inspector, Henry "has at all times cautioned the Inspector not to impose any unnecessary hardships on the operator."[59] Elsewhere, chief inspectors approached heroic dimensions. Illinois miners could on the one hand assert that the Illinois inspection service "is not giving satisfaction to the mine workers" and on the other seek the directorship of the Bureau of

Mines for their chief inspector and former UMWA organizer David Ross. In Pennsylvania, Ohio, and Kentucky, the chief inspectors played leadership roles, riding herd on recalcitrant district inspectors and insisting on thorough and impartial enforcement. Responding to criticism that his inspectors were not forcing compliance with the law, Pennsylvania's Roderick once transferred half of his inspection force to new locations. J. M. Roan, who as safety commissioner was responsible for Ohio's mine inspection after 1914, was typical of the zealous chief inspector. Roan was firm with operators seeking special treatment. When one operator failed to provide a second opening as instructed by the district inspector, Roan suggested that he do so and "avoid any unpleasantness in the matter."[60] A request from another operator to maintain only one opening was firmly turned down. A coal company which had neglected to forward fire-boss reports to the state received a standard hard-line letter from Roan: "You might look upon this matter lightly, but the Mining Department considers it a very serious proposition."[61] Like Roderick, Roan worked to perfect the state's enforcement mechanisms. To cope with the hazards of dry winter dust, he established a system of double inspection, under which a district's regular inspector was joined in his rounds by an inspector from another district. This new system also meant unannounced inspections, for Roan affirmed that mine managements "will have to attend to their 'knitting' or there is liable to someone drop in on them at any time." [*sic*][62] To eliminate excuses offered by operating companies for noncompliance with inspector orders, Roan worked with district inspectors to develop a uniform time-compliance system.[63]

Evidence of insufficient diligence on the part of district inspectors is of two varieties. Miner charges of inspector neglect were common. "At present a law means little and a mining rule less," wrote the *United Mine Workers Journal* following explosions at Pocahontas, Virginia, and Raton, New Mexico, in 1906. "The inspectors do not enforce the one and mine bosses disregard the other."[64] Although the *Journal* usually argued that state legislation was inadequate, after the 1902 Fraterville, Tennessee, explosion it made the point that "the mining laws were sufficiently broad to compel obedience to the recommendations of the inspector if he were inspired by a proper sense of duty."[65] Miners also accused inspectors of ignoring their obligations to pit committees (groups of miners seeking a voice in mine management) and of being too lenient with operators. *Coal and Coke Operator* confirmed the charge of leniency when it noted that the electricity provisions of the 1893 Pennsylvania mine law have been "regarded as a dead letter by the mining department, more or less . . . because it was recognized that to [enforce them] would interfere with steady and profitable employment for many thousands of men and millions of capital."[66] The second variety of evidence is of an entirely different nature, amounting to a logical deduction from observable patterns in the inspectors' reports. Although most inspectors encountered unsafe conditions in their districts and/or some resistance to enforcement on the part of miners or operators, there are cases of inspectors reporting virtually complete satisfaction with conditions and attitudes. Lance B. Holliday of West Virginia's Ninth District, for example, in 1910 reported "a universal tendency on the part of operators and officials to fully comply with the Mining Laws in every respect." R. S. LaRue of the First District claimed he had "no difficulty in having the mining laws complied with this year." One step down from such all-encompassing statements were comments like that made by I. M. Kelley of the First District, who admitted encountering some problems in ventilation and drainage when he took over the post but held that the defects had been remedied. "My suggestions," he said, "met with a hearty co-operation from all in authority."[67] If such statements were present in every inspector's report, we could then either judge the entire inspection force incompetent or conclude that these words, however absolute they seemed, couched some kind of relative standard. In fact, since other inspectors experienced difficulty and reported it, and since the districts of Holliday, LaRue, and Kelley were of the same character, we can only conclude that the positive reports of these inspectors and others like them served only to mask reality and deter enforcement of the mine laws. With the criticisms of mine inspectors, these saccharin reports indicate that a minority of state inspectors—mostly from West Virginia—did not enforce the mine-safety laws.

The inspectors, of course, were only one element in the enforcement equation. Miners, operators, and mine management (superintendent, mine manager, boss, fire boss) were also crucial to the result, and there is evidence that each of these groups met its safety responsibilities less than fully. While the utility of blaming the miner for accidents can be questioned, and although criticism of miners served purposes which went well beyond logical analysis, it is nonetheless clear that miners continually violated the safety laws and often resisted the orders and suggestions of inspectors. To some extent, the violations varied with state legislation. In West Virginia and Pennsylvania, miners ignored regulations against solid-shooting; in Illinois, they regularly transgressed contract provisions by using too much powder to shoot the coal. Everywhere, to judge from inspectors' reports, they rode illegally on mine cars, carried too much powder into the mines, used the wrong oils in their lamps, failed to timber properly their working places, and were careless in handling explosives. Two brothers died in Jefferson County, Ohio, when a spark from a lighted lamp ignited a keg of powder being used as a seat.[68] Almost without exception, state inspectors cast the miners in a central role in mine safety. Ohio's Roan scored old miners, who "seem to think it is cutting their pride, more or less, to accept instructions."[69] In three widely separated reports, Roderick of Pennsylvania emphasized that miners presented more serious obstacles to enforcement than operators. "I have found it a rare thing," he said, "for an operator to refuse to comply with any instructions or suggestions given by the inspectors regarding the preventing of accidents. If the employes gave the same kindly consideration to suggestions from the inspectors that the operators give, fewer accidents would have to be recorded."[70]

Roderick's comments notwithstanding, owners and managers were not always, or even usually, the willing servants of the law. West Virginia Governor William Dawson was probably closer to the truth when he divided coal operators into two classes—those "who yield cheerful obedience to the law, who see in the inspector a friend and an aid, and consider the bureau of mines an institution for their benefit; [and those who] resist the enforcement of law, see in every inspector a spy and an enemy, and look upon the bureau as an invention of oppression."[71] Though there is evidence that the first category grew rapidly following the bad years from 1905 to 1908, the second category, or variations on it, was always prominent. Neither miners nor inspectors, after all, could maintain proper ventilation equipment or currents, construct fireproof underground stable areas, install shields for mine machinery, provide sufficient supervision, or fulfill any number of other management functions. For every miner who initiated an explosion by shooting coal off the solid, there was a mine foreman who failed to measure the air currents or instruct the miners in proper technique, or an operator who allowed the dust in his mines to dry and accumulate. For every miner who died because he chose not to timber his working place, another died because a foreman had not provided him with timbers. Operators neglected to provide up-to-date mine maps, failed to post special rules for government and operation of the mines, sold illegal oils to their employees, and insisted on employing mining machinery even when it led directly to promiscuous shooting. Although the necessity of good ventilation had been recognized since 1870, operators were particularly negligent in this critical area.[72] Frank Parsons, inspector in West Virginia's Second District, noted that the problem of ventilation apparatus—doors, overcasts, stoppings—was "the worst thing with which I have had to contend."[73] Another inspector found poor ventilation in many operations and traced the cause to incompetence. Foremen were measuring the air currents, he said, but at the wrong place. The extent of the problem is revealed in a 1900 survey conducted by the West Virginia inspectors. Of 320 mines inspected in that year, eighty-six received a rating of "fair," eighteen were rated "bad," and the remainder were classified as in "good" condition. Virtually every mine rated in fair or bad condition—almost one-third of the total—was listed as deficient in ventilation.[74]

The policies that produced such conditions were generally sustained by mid-level management. Hired to get out the coal and make the mine a paying proposition, superintendents, foremen, and even fire bosses took this, rather than safety, as their primary function. In one case, a mine manager resisted changes that would increase the cost of production even when the changes had been approved by the owners. Similar reasoning led a Pennsylvania fire boss to allow the men to enter the mine even after he had located dangerous gas concentrations. Illinois miners claimed to have knowledge of fire bosses who conducted inadequate examinations because they were afraid of being discharged by the company. Mine inspectors' reports confirm the literary observations of Powers Hapgood: mine managers were often incompetent and almost always closely tied to the owners' interest in production.[75]

Hostility and a strange kind of gamesmanship characterized operator-inspector relationships in a number of cases. Compliance with inspector orders was often half-hearted, incomplete, or slow, a tactic rather than a genuine attempt to fulfill the law. Thomas Waters, inspector for Ohio's First District, wrote of operators who were little affected when an inspector closed their mines. They made a few improvements—enough to be permitted to resume work—and then turned their whole attention to production, ignoring health and safety.

West Virginia's J. G. Boyd found most operators inclined to cooperate, but was deeply distressed by "a few who pretended that they were in sympathy with the department," but who, when put to the test, were found to be "insincere in their convictions."[76] Inspectors had only a limited number of weapons with which to counter this negativism of operators, mine officials, and miners. The state mine-safety statutes prescribed fines for violations; they were seldom large, even when a corporation was involved. Inspectors could not assess fines on their own authority; only the courts could do this, following a conviction for violation of the law. All violations were misdemeanors rather than felonies. If death occurred, coroners' juries were responsible for initiating criminal proceedings. Besides prosecuting, the inspector could, through a variety of processes worked out in the state legislation, close a mine he thought unsafe. In most cases the law required an inspector to accomplish this by securing an injunction through the proper court. Finally, inspectors in some states possessed the power of publicity—the ability to shame recalcitrant operators into obedience by publishing accounts of fatal accidents.

Of these tools, only prosecution was of much use. Inspectors closed mines in Illinois, Kentucky, West Virginia, Ohio, and perhaps other states, but they did so infrequently and received as much opposition as aid from local courts. Ohio operators became incensed when chief inspector Harrison liberally interpreted the state law to allow inspectors to close mines at their discretion. The Ohio Coal Operators' Association claimed that since the law expressly provided that inspection "shall not unnecessarily obstruct the operation of mines," the power of inspectors was limited to specific orders. If enforcement procedures were as the operators said, wrote Harrison, then violators could simply discontinue the practice in the inspector's presence and wait for his return. He concluded:

> Every law on the statute book might be ignored with the greatest impunity, and without fear of prosecution.
>
> We have heard some very peculiar opinions regarding the Ohio mining laws, but we trust you will pardon us for saying that the one contained in your letter is without a doubt the most unique, far-fetched and utopian construction we ever knew applied to them. At first, we were disposed to treat this definition as a huge joke, perpetrated on the two writers of the letter or on the chief inspector of mines, by some witty humorist or willful wag who was more encumbered with surplus time than onerous duties. . . .

> We feel sure, however, that there is not five per cent of the coal producers of Ohio, that would want to have their mines operated under any such loose system as your definition of the law would inaugurate.[77]

In West Virginia, inspectors applied the injunctive remedy with some zeal after 1906, but they soon discovered that coal companies had the advantage in presenting evidence to the courts. In 1908, chief inspector Paul was angered when a court reopened the Jed Mine in McDowell County, which the mines department had closed because of dangerous gas accumulations, and then rejected a petition for appeal.[78]

Coroners' juries also contributed little to the enforcement process. Governor Dawson of West Virginia captured the essence of the institution in a report to the legislature on the recent Pocahontas disaster. The coroner's jury handling Pocahontas, said Dawson, had "rendered a verdict that the cause of the explosion was not known, but which declared none of the mine people blameworthy. Just how the jury could arrive at that result is a little puzzling. If the cause of the disaster was not known it would seem impossible to know that no one was to blame for it. . . . The fact is that investigation by coroner's juries into mine disasters rarely amount to anything." [*sic*][79] Not all coroners' juries fit Dawson's description. There were a number of cases in which coroners' juries recommended grand-jury proceedings against miners, inspectors, foremen, superintendents, and even owners; and occasionally a coroner would recommend changes in the mine laws. On the whole, however, activist coroners' juries were regarded with wonder by contemporaries. The institution served mainly to prejudice law-enforcement officials against legal action and to shift the burden of prosecution to district attorneys.[80]

In 1906 Ohio's Harrison, convinced of the need for stronger enforcement, ordered his inspectors to threaten mine bosses and miners with prosecution, and within two to three years Ohio, Pennsylvania, West Virginia, and presumably other states, had all turned in earnest to the courts. Prosecutions increased dramatically, from a smattering before 1905 to more than 150 in Pennsylvania and West Virginia in 1910. But the movement was sheathed in limitations. Everywhere fines were minimal, even for superintendents and foremen. Although Harrison claimed that the strongest feature of the new 1910 mining law was its penalty provisions, in 1911 fines collected under the act amounted to only $400. There, as elsewhere, imprisonment for safety-law violations was unheard of, even, apparently, for those responsible for fatal accidents.[81] The reports of the state mine inspectors, although probably an incomplete source of information, provide ample evidence of selective enforcement and class-biased justice. To begin with, the reports for Pennsylvania, West Virginia, and Ohio contain little evidence of prosecutions before 1904 and none after 1912, reflective of the limited influence of the safety movement except within a narrow chronology. By any standard—recorded violations, numbers of mines and employees, the detailed and technical nature of mine-safety legislation—the total numbers of prosecutions were minimal. For the three-state area in 1910, prosecutions totaled 395; for 1911, 312—and these clearly were peak years. Moreover, law enforcement officials were extremely reluctant to prosecute higher management. In West Virginia and Ohio, miners and mine bosses were virtually the only ones prosecuted, the former primarily for solid-shooting, the latter usually for failure to measure air currents. Of a total of 163 prosecutions in West Virginia in 1910, miners accounted for 159. Pennsylvania completed 489 prosecutions in the four years from 1908 through 1911. Of these, 27 were of superintendents, 39 were of foremen, 31 were of fire bosses, and the remainder, 392, were of ordinary mine workers.[82]

Selective enforcement was perhaps abetted by xenophobia. Writing to the Industrial Commission, safety commissioner Roan of Ohio emphasized the need for inspectors to have the power to make arrests of miners immediately upon observation of a violation. This was, he said, especially urgent in the case of foreigners: "many of them are like white eggs—when you mix them up in a dozen you cannot tell which one you had in your hand to start with."[83] West Virginia inspectors, moreover, evidently ceased prosecuting operators and managers when it became clear that they could not be convicted. "Through some process not clear to this department," said chief inspector Paul, "the courts have not disposed of prosecutions pending against mine officials."[84] A district inspector reported that workers had "completely lost all confidence in the local courts . . . [and were] thoroughly convinced that justice could not be obtained towards the enforcement of the mining laws."[85] Only for Pennsylvania is there evidence of prosecution of superintendents, and there, too, the great majority of cases were brought against mine workers. The inspector reports provide no proof that mine owners were prosecuted in any state. Major journals of operator opinion applauded and encouraged selective prosecution, which merged conveniently with their view that miners were responsible for safety problems.[86]

With the enactment of Pennsylvania's 1911 mine-safety law, an era of frenetic state activity came to an end. The next decade would bring only refinements. From the moment this impressive legislative structure was created, however, it was outmoded, a nineteenth-century solution to twentieth-century problems. An industry serving national markets, shipping its products hundreds and thousands of miles across state borders and employing a national labor supply which refused to remain stationary, had constructed its safety edifice on firm but inappropriate state foundations. It is difficult, of course, to conceive of its having been done any other way. Industrial safety was a natural state obligation, a historically sanctioned function of the police power of the states. Few voices were heard proclaiming the benefits of national social legislation except in industries such as railroads and the merchant marine that were widely recognized as interstate. When the winter of 1907–1908 made a mockery of existing state legislation, the reformers did not drop their aversion to national safety legislation. Instead, they sought to nationalize the existing legislation, first by creating a national Bureau of Mines to furnish state legislation with a scientific underpinning; and second, through a campaign to make mine-safety legislation uniform from state to state. Uniformity, and a related concept, standardization, at once asserted the negative view that national legislation in coal-mining safety was constitutionally inappropriate and the positive view that some form of national action was necessary. In short, the idea of uniform state legislation expressed an ambivalence built into federalism and was designed to bridge a temporary gap in the constitutional system.

Uniform state legislation and standardization were considered desirable goals in many areas besides coal-mining safety, most commonly where safety was a factor. The Industrial Commission in 1900 recommended, for example, that a uniform law regulating hours of labor for industrial occupations be adopted by all the states. Almost two decades later, the Bureau of Standards, the Department of Commerce, and the Working Conditions Service in the Department of Labor began investigations which were designed to lead to uniform safety codes for building equipment, fire prevention, elevators, cranes and derricks, conveyors, steam engines, oil and gas engines, textile machinery, explosives, and rubber goods. The movement for uniform state legislation attracted A. Mitchell Palmer, who urged uniform child-labor legislation, and produced a conference on the general topic, sponsored in January 1910 by the National Civic Federation. Like Palmer's desire for uniform state legislation, that of the National Civic Federation arose from an essentially conservative base, from an opposition to further centralization of power. Among the organizations attending the conference were the Grange, the AF of L, the National Association of Life Insurance Presidents, the AMA, the Census Bureau, the American Public Health Association, and the National Association of State Labor Commissioners. Others promoted uniform legislation to cope with the problems of habit-forming drugs, divorce laws, corporation laws, direct election of senators, and quarantine.[87]

Miners, operators, engineers, inspectors, and government officials were nearly unanimous in their advocacy of uniform coal-mining safety legislation. From the engineering standpoint, uniform legislation was feasible and desirable for its effect in simplifying statutes. "The engineer," noted J. A. Garcia from personal experience, "must either stuff his brain with a mass of legal lore and keep stuffing afresh as new legislation is enacted, or make his own code and index the laws of each state for convenient reference."[88] Engineers agreed that the varied conditions under which bituminous coal mining took place presented no important obstacle to uniform safety legislation. More common was the argument that uniform legislation was necessary because of the migratory nature of mine labor. Miners, said Pittsburgh mining engineer W. E. Fohl, were of low intelligence and were natural transients. He suggested uniform rules and laws "in order to firmly fix their observance in the habits of this floating contingent."[89] Drawing an analogy between miners and ducks, Iowa inspector William E. Holland said: "A large portion of the coal-mining fraternity, like the wild ducks, migrate with every change of season. . . . They find a different set of laws in every state and accidents are liable to happen from unfamiliarity with the law in the state to which they have removed. With a uniform code, this condition would be eliminated."[90] The bureau's O. P. Hood voiced the same concern in 1916 when he recommended a uniform signaling system: "As it is now, when a miner goes from one camp to another he has a set of signals to learn and if confused may give the wrong signal and endanger life."[91]

Uniformity was most often defended on the grounds that economic conditions in the bituminous industry made mine-safety legislation an economic liability unless similar legislation were passed in competitive states. The coal industry (i.e., the central competitive field), the argument went, was so intensely competitive that legislation passed in one state put that state at a competitive disadvantage until similar legislation was passed in other states in the field. Two operators, from opposite ends of the central competitive field and both interested in uniform workmen's compensation legislation, made this point with some clarity. Z. Taylor Vinson put forward the case of the West Virginia operator: "If we should have a workmen's compensation law in West Virginia, and you do not have it in Pennsylvania or Illinois, or other coal producing states, then obviously the coal operators of West Virginia will have to bear that burden alone, and it will be impossible under competitive market conditions to bring about any uniformity unless all the states adopt substantially the same plan, but if it exists in practically the same degree throughout the whole country, then the burden would be borne by all in the first instance and ultimately refunded by the consumer."[92] Operator A. J. Moorshead of Illinois, one of the foremost advocates of uniformity, described Illinois's workmen's compensation act as "a good measure if it were a national one," but burdensome "when it is saddled upon one State and not upon the neighboring states."[93]

Though a favorite among operators, the argument from economics was widely recognized as valid. John Mitchell, president of the United Mine Workers for a number of years, was skeptical of the possibility of achieving uniformity in legislation but still accepted industry economics as uniformity's primary justification. Even the *United Mine Workers Journal*, admittedly interested in uniformity as a means of upgrading legislation in states where the union was weak and without political influence, purported to understand the economics of competition. "We prefer to believe," stated the *Journal*, "that most of the operators would prefer the general adoption of safeguards, providing their competitors were also obligated to adopt the same."[94] Such important figures in the uniformity movement as Governor E. F. Dunne of Illinois, James Roderick, Van H. Manning, and J. T. Beard of *Coal Age* might have seconded engineer Fohl's suggestion for a new commandment: "Thou shalt not sell coal at a lower price than thy neighbor if such selling is only possible by reason of neglect of the natural laws which govern the preservation of human life and the conservation of Nature's bounty."[95] In practice the ingredients of a competitive advantage encompassed more than safety legislation, so that, for example, the nature of West Virginia's coal seams and labor supply would have provided that state with a competitive advantage no matter what state legislation was passed. Nonetheless, the argument which tied safety to competition was so widely accepted and so seldom questioned that its truth was a matter of little practical importance. Liberals treated it with respect. In a piece written for the *Survey* in 1909, Charles McCarthy of the Wisconsin Legislative Reference Library referred to the fear of interstate competition as "a bugaboo." He insisted, however, that Wisconsin businessmen would not accept that view until they had been provided with data and information. John B. Andrews, secretary of the American Association for Labor Legislation, also expressed the power of the competitive argument. "Too much emphasis," he said, "has sometimes been put on the interstate competition argument in protective legislation, but there is a real problem here that calls for uniformity of minimum legal standards."[96]

Uniformity of state legislation was not, of course, the only possible solution to the peculiar safety problems arising from an interstate coal industry with nearly national markets and a migratory labor supply. Federal regulation, by commission or Congress, also had its advocates. Among them was a senator, who introduced a bill calling for federal inspection and regulation of all coal mines, and a congressman, more conservative, desiring federal regulation as the only road to uniformity but believing that the federal Constitution prohibited the requisite legislation. Moorshead for a time saw federal control of the entire mining industry as the only solution.[97] Other leading proponents of federal regulation for the coal mines included John R. Haynes, California's commissioner of mines, who viewed the current state regulatory system as inefficient and proposed instead a five-man commission analogous to the Interstate Commerce Commission, and Van Amberg Bittner, who occupied a number of positions in the Pennsylvania hierarchy of the United Mine Workers. In 1916 Bittner's views became official UMWA policy when the national convention declared itself in favor of ownership and control of the coal industry by the federal government.[98] Federal regulation was not, however, a real alternative to uniformity and standardization in the Progressive period. Bureau supporters James R. Garfield, Manning, and Holmes all rejected federal regulation in favor of uniformity. Holmes did not see the mines as properties of interstate commerce and said that he hoped any movement for federal regulation or rule-making "won't meet with much success because I think that ought to be cared for by the state."[99]

A number of advocates of uniform legislation indicated a willingness to go beyond uniform legislation to some kind of truly national action if the movement failed. This is of some importance, since it indicates that uniformity was, for some, as much a transitional concept looking forward into national legislation as an attempt to support the old state system. The people, said the *Outlook* in a 1913 editorial, "are coming to realize that unless the States co-operate to secure legal conformity the people will look to the Federal Government for relief, an event that would foreshadow the withering of State power."[100] The next year the convention of the United Mine Workers was the scene of an important debate on whether the states or the federal government should have jurisdiction over safety and other mining matters. A number of delegates who had become soured on uniform legislation led the convention to support a resolution calling for more activity in behalf of uniform legislation only if an appeal to the Congress failed. President John White's insistence that national legislation would require an amendment to the federal Constitution seemed to deter the delegates not at all, for they were convinced that uniformity would not work in states like West Virginia and Colorado, where the operators controlled the legislatures. "It would be useless," said one delegate, "for us to try to get laws like that on the statute books. Why couldn' [*sic*] we get it before the United States Government and get some action on it? The Federal Government controls the whisky question from one State to another."[101] There is also something instructive in the model used by uniformity proponents—the national, unitary legislative and inspection systems of Britain, France, Belgium, and Germany.[102]

The legal profession, particularly, was aware of the substantial legal and constitutional importance of the movement for uniform state legislation. In 1916, the report of the American Bar Association Committee on Uniform State Laws claimed that "uniformity is not simply a name, it is a principle, and a principle which is of the very essence of democracy, if we mean by democracy that state of society in which there is one law equable in its application to the rights of all men alike everywhere."[103] Although proponents of uniformity commonly defended the practice as one which would maintain the Constitution intact, it was also widely recognized that uniformity was a means of nationalizing a federal constitutional structure. "The Commissioners on Uniform State Laws," wrote one observer, "is an extra-constitutional body designed to accomplish the very thing that the Constitution would prevent, or at least that which the Constitution directs should be done in a different way."[104] Ostensibly only a voluntary program of state action, uniformity was for some a stalking horse for nationalization.

The migratory nature of labor and the competitive qualities of the market in the coal industry thus combined with a strong sense of resistance to direct federal regulation to produce a climate in which uniformity was pursued as possible and desirable. Within this framework, a national scientific organization was the specific, if intermediate, goal. In 1906 John Mitchell supported a bill to create a bureau of mines, suggesting that the encouragement of uniform protective legislation would be one of the proposed agency's five principal functions.[105] Two years later, when Congress came close to creating a bureau, the potential impact of the institution on state safety legislation was a common supporting rationale. In one of his more eloquent moments, Ohio inspector Harrison claimed that "absence of an established central body is a 'missing link,' an immovable stumbling block in the way of effective concert of action by various state mining departments."[106] While operators were more cautious lest a federal agency interfere with state inspection functions, one operator journal ventured to suggest that a new bureau might bring about uniformity through its "influence and prestige."[107] Holmes, as bureau director, incorporated the uniformity idea into his four-part bureau plan of action. Under part one the bureau made mine-safety investigations and disseminated results; under part two (uniformity) the states enacted appropriate legislation for inspection and safety. For the bureau, too, uncoordinated state legislation was not enough; federal legislation was too much. Uniformity promised the results of federal legislation without its drawbacks. Unfortunately, the bureau was ill prepared to assume much responsibility in the movement for uniform state legislation. In many cases it had no knowledge upon which to base recommendations to the states or had not compiled relevant codes and statistics. Sensitive to criticism for interfering in state concerns, Holmes and Manning steered their agency clear of state legislative matters, though the bureau followed state legislative movements, particularly in Illinois and Pennsylvania.[108]

Only in two areas—mine accident statistics and electricity—could the bureau's involvement in matters of uniformity and standardization be considered even moderately successful. The need for standardization of mine-accident statistics arose because of disparate criteria employed by the states in classifying mine accidents and determining fiscal years and subjects of study. Some states, for example, had records of deaths in their mines but none of accidents resulting in injuries only. When the Bureau of Mines first became involved in statistical standardization in 1910, statistics were collected from operators, state-mining departments, and state inspectors. Under James Paul and Edward W. Parker, the bureau within two years had strengthened the relationship with the state mine inspectors and was relying solely on them for its statistical information. The first monthly report of coal-mine accidents was published by the bureau for discussion at the annual meeting of the Mine Inspectors' Institute of America (MIIA) in June 1911. By 1913 every state except four—Arkansas, Oklahoma, Kentucky, and Illinois (and arrangements had been made with Illinois)—was cooperating with the bureau, and when Holmes made public his dissatisfaction with the lagging states, he soon had their cooperation. A national conference, called by the bureau for February 1916 and attended by mine inspectors, labor commissioners, chairmen of mine-rescue commissions, state industrial accident board members, statisticians, and representatives of compensation commissions, brought further refinement to accident standardization and produced a committee to recommend to the states standard forms for mine statistics. While experts could still find substantial fault with the system—use of the tonnage basis (accidents per ton of coal mined) and a dearth of data for nonfatal accidents were frequent complaints—a solid beginning had been made.[109]

The first organization to work in the area of electrical standardization was the American Mining Congress. Its Committee on the Standardization of Electrical Equipment in Mines, established in 1908, within one year had drafted a code of electrical rules for enactment into law by state legislatures and, perhaps a victim of its own success, was discharged after making its first report. The code developed by the committee was adopted almost verbatim by the Pennsylvania legislature in 1911. After 1912, the AIEE, the AIME, and the MIIA became involved with the bureau in work on electrical codes. Two other organizations—the National Safety Council and the United States Bureau of Standards—were rebuked for involvement in an area which, according to the bureau, was not in their province. Most of the efforts at standardization of electrical codes went unrewarded. The Pennsylvania experience was not repeated, although a few states did enact stricter laws for electrical equipment based on bureau recommendations. As late as 1920 the Pennsylvania anthracite field and many bituminous states had few electrical regulations, and the MIIA was having its first full-scale debate on electrical standardization. Sorely disappointed with progress in electrical standardization and particularly disturbed by the obstructionist tactics of operators on the MIIA's legislative committee, Illinois mine inspector Joseph Haskins was ready to turn to federal legislation. "While most of us know little of the principles of electricity," he told the 1920 convention of the MIIA, "we do know the mining game and our experience has been such that we have little hope of obtaining what we need in the way of legislation other than through the national government."[110]

Although they could claim no more success than the Bureau of Mines, the Mine Inspectors' Institute, the American Mining Congress, and the state of Illinois were more active in pursuit of uniform state legislation. Organization in pursuit of uniform coal-mining safety legislation began in response to the explosions of December 1907, when the editor of *Coal Age*, James T. Beard of Scranton, Pennsylvania, acting at the request of chief inspectors George Harrison, James Roderick, and James Paul, plumbed sentiment for a national organization of inspectors. At the first meeting of the Mine Inspectors' Institute in June 1908, the inspectors affirmed their interest in uniform legislation and moved on to substantive matters. Recognizing the importance of scientific investigation to uniform legislation and standardization, the institute under the presidency of Harrison, who served from 1908 through 1911, lent strong support to the campaigns for a national bureau of mines and for additional testing stations. Signs of difficulty first appeared at the 1910 meeting, when the Committee on Uniformity, created in 1909 to prepare a list of substantive provisions to be embodied in state mining codes, announced that it had been unable to meet until the week of the convention. The committee's hastily prepared report proved too vague to be useful to state legislators. Although one committee member, Peter Hanraty of Oklahoma, presented a list of sixteen rules for consideration (six dealing with shotfiring), the committee apparently thought them inappropriate and they were not made part of the report. It is probable that even at this early date there was division within the institute over precisely what matters should be the subject of uniform legislation.[111] Having missed the opportunity to influence the crucial legislative packages produced in Ohio, Pennsylvania, and Illinois in 1910 and 1911, the MIIA moved into a decade of apathy and impotence. From 1911 through 1914 the inspectors showed little interest in the subject, even in 1913 when a strong advocate of uniformity, Thomas K. Adams, held the institute's presidency. Beginning with the 1914 meeting, a partial revival of interest in uniformity occurred, led by the institute's organizer, J. T. Beard. As chairman of the Resolutions Committee, Beard indicated his desire to awaken the institute by announcing that resolutions should in the future be more practical than they had been previously, and a year later he was openly critical of the institute for its general inactivity and insufficient attention to uniform legislation. In answer to the multiplying critics of uniformity, Beard admitted that uniformity could not be applied to all details of coal-mining safety, but he was adamant that certain subjects—particularly certification of mine officials and rules for classifying the mines with respect to dust, gas, and ventilation—were appropriate subjects for uniform legislation. A simple resolution calling for competency examinations for all positions of authority underground were the firstfruits of Beard's labor and the first substantive commitment of the mine inspectors in several years. By 1919 his committee had renewed its rather meager efforts with a series of resolutions dealing with explosives.[112]

That an organization formed expressly to pursue uniform mine-safety legislation could so quickly and completely lose interest in its goal requires some explanation. Poor timing is part of the answer. When the group finally was capable of action, the states had completed a major legislative cycle. There was comparatively little lawmaking between 1911 and 1920, and the institute was left with little to do. Moreover, an increasingly large element in the institute had begun to see immense difficulties in preparing a code of laws acceptable to all or even most of the coal-mining states and therefore viewed uniformity as impractical or as a violation of states' rights. Roderick of Pennsylvania, who as president had advocated a more active role for the MIIA, by 1915 could only lament that uniform mine legislation was an impossibility due to diverse conditions in the states. Discussion of uniform legislation was a waste of time since only a few things could be made uniform. "I believe," he said, "that the mining laws of every State must deal with its own conditions."[113] Support for Roderick's position came from R. A. Shiflett and James Dalrymple, chief inspectors of Tennessee and Colorado. Others, including P. J. Moore of Pennsylvania, who questioned whether the institute had "any right to presume to advocate a general law," and J. J. Rutledge of Oklahoma ("I do not believe we can make any hard and fast rules in coal mining"), emphasized that the complexities of coal mining necessitated administrative, rather than legislative, regulation.[114]

Organizationally, the inspectors were hampered by a number of factors incident to the occupation and its politics. Inspectors were divided politically: some state delegations were dominated by miners, others were closely tied to operator interests. Within the national organization continuity was a persistent problem because mine department personnel changed frequently and the states sometimes refused to cooperate in allowing inspectors to attend conventions. Perhaps the problem lay in the personality of the mine inspector, described as "a conservative man . . . a man that says very little in public, and . . . is very cautious about what he does say."[115] Whatever the cause, the mine inspectors played their role in the movement for uniform mine legislation with little enthusiasm. For an organization formed to further uniformity, the accomplishments of the Mine Inspectors' Institute were paltry indeed.

The Illinois movement for uniform legislation bore little similarity, in personnel or procedure, to the efforts of the mine inspectors. It was initiated by Illinois operators caught in a competitive squeeze and looking to uniformity to spread the burdens of safety legislation throughout the central competitive field. "The way we are going now," said A. J. Moorshead, president of the Madison Coal Corporation and one of the movement's leaders, "some States have little or no legislation to govern them in mining, while others are saddled so heavily that they are brought to insolvency, which, as I said before, is the case in Illinois."[116] As early as 1911 Governor Charles S. Deneen publicly recognized the importance of uniform legislation in the coal-producing states and suggested an interstate conference. No action was taken until the Illinois Mining Investigation Commission entered the picture in 1915, requesting Governor E. F. Dunne to invite representatives of coal-mining states to form an interstate commission for the drafting of uniform mine legislation. Necessary mine safety and conservation legislation, said the Illinois Commission, could not be enacted in Illinois precisely because of competitive conditions: "Under existing conditions in some of the states, the industry actually enjoys a premium for unsafe conditions in comparison with the men engaged in the mining industry in the State of Illinois. These are evils directly flowing from a want of uniform legislation governing the coal mining industry."[117] Dunne sent state governors an invitation to appoint representatives to meet in Illinois in conjunction with the American Mining Congress's annual convention. In the invitation he noted the benefits of uniform legislation in commercial areas. "No competition between the different states of the Union," he said, "should be allowed to exist based upon the cheapness of production at the cost of human life and limb, and which tends to discourage progressive legislation; no premium should be placed upon the financial conditions of recklessness in mining methods." At the Bureau of Mines, Van H. Manning applauded the fledgling movement.[118]

For a time the Illinois movement appeared likely to evoke national action. Operators, miners, engineers, inspectors, and others who had been appointed to state commissions by their governors met in Chicago in November 1916 and reconstituted themselves as the Uniform Coal Mining Laws Association, with Moorshead and Robert Harlan of the UMWA as president and vice president. Using the Illinois Mining Investigation Commission as a model of group compromise, this new association (essentially a section of the American Mining Congress) called for the creation of a nine-member Commission on Uniform Coal Mining Legislation, to consist of three miners, three operators, and three representatives of the general public with the director of the Bureau of Mines an ex-officio member. This commission was to draft and present to the next annual meeting of the Mining Congress a code of uniform coal-mining laws which the Uniform Laws Association could then recommend to the legislatures of the states. Unfortunately, nothing was done to pursue these plans in the following years. The Resolutions Committee of the American Mining Congress rejected the proposed Commission on Uniform Coal Mining Legislation and the AMC took no further action. High prices caused by the extraordinary demands of a war economy temporarily vitiated the economic impulse which was at the base of the Illinois movement. Along with other relatively high-cost producers, Illinois coal operators no longer had their competitive relations determined by the lowest common denominator—usually West Virginia production. Five years earlier, when interstate competition had made uniform legislation logical and necessary, operators had been so divided that the degree of cooperation requisite to uniformity had proved unachievable. "One state does not," Moorshead had noticed, "in the matter of legislation, care a continental about the others."[119]

"We live in an age of organization," said the president of the Mine Inspectors' Institute in 1913.[120] Mere knowledge of this truth, however, was not enough. Failure marked the various movements for uniform coal-mining safety legislation precisely because the organizational matrix brought to bear on the problem—the Mine Inspectors' Institute, the American Mining Congress, the Bureau of Mines—proved unable to bring cohesion to a fragmented and archaic political system and was incapable of resolving the contradiction inherent in a national industry regulated by state laws. To solve its problem the coal industry would eventually turn to national legislation.

CHAPTER

4

The Miner and the Union

THE MINER worked in a highly competitive, labor-intensive industry, in which operators routinely met their competition by reducing the wages of the working man. Whether they did so directly, by paying him less for the same work, or indirectly, by increasing prices at the company store or establishing a new system for measuring miner production, made little difference. In either case, the miner was under pressure to maintain his standard of living, even if that meant employing unsafe mining techniques, and his ability to respond to this pressure by moving into another line of work was considerably circumscribed by the geographical isolation of most bituminous mining operations. This isolation also made the miner dependent on the operator for much of his existence and correspondingly diminished his willingness to protest against unsafe working conditions. If one turns from the roles of employee and wage earner to the role of the miner as worker, it becomes clear that an image of the miner as economically motivated and dependent is valuable but limiting. As a worker, the miner was unusually independent. He worked alone or with one other miner, usually without supervision or even instruction. Experienced miners, in particular, came to see this independence as a valuable right and to resent infringements upon it, a self-concept which came into direct conflict with that aspect of the coal-mining safety movement which defined progress as increased supervision of the miner at the working face. Finally, great numbers of coal miners in this period were recent immigrants, largely without a speaking knowledge of English; they, too, contributed to the prevalent and deserved reputation of the miner as a careless employee. Much of the Progressive movement in coal-mining safety was an attempt to reduce the extent of that carelessness and its consequences. This analysis must, of course, be interpreted within the context of the entire mine-safety problem. The miner was only one element in that problem, and not the most important one. Whether a miner employs an unsafe technique is, after all, not absolute but relative to the condition of the mine.

Miners in many areas were members of a strong national organization, the United Mine Workers of America. Through its national, district, and local organizations the UMWA worked for state and federal legislation and made more limited contributions to the coal-mining safety movement through use of the contract and, when it seemed necessary, through the strike. However, the union in this period was interested mainly in membership, and its membership was interested primarily in economic issues, not in safety. Some union demands were, in fact, antithetical to the requirements of safety, and the union sometimes used the contract and the strike to achieve goals which, while justifiable in the union context, were roadblocks for mining safety.

In the hands of operators, miner carelessness became less a keen analysis of the causes of mine accidents than a rationale or tactic. As mine disasters multiplied and increased the pressures for new legislation, operators affirmed the excellence of existing law and used the issue of miner carelessness to shift the locus of the discussion away from additions to the statutes. Mine managers protested that they should not be held responsible for accidents, since they could not possibly supervise every careless or incompetent miner. The *Coal Trade Bulletin* epitomized this view of mine safety. By 1905 it had concluded, without data, that 99 percent of all mine accidents "are due absolutely to the carlessness or wilful negligence of the men employed in them" [*sic*]. While the Monongah explosion affirmed for many the need for scientific research into the causes of mine explosions, the *Bulletin* found its own lesson. "The human element in the industrial equation," it noted, "is the one unknown quantity that sets at naught all efforts to safeguard properly the lives of the workers in mine or mill. This fact was never more clearly demonstrated than in the disaster which snuffed out so many lives at Monongah." When Pennsylvania's chief inspector James Roderick maintained in an annual report that carelessness was responsible for more mine accidents than any other single factor, *Coal Trade Bulletin* said this expert testimony demonstrated "beyond cavil, that legislation cannot remedy the evils which result from the perversity of human nature." The Cherry mine fire was analyzed along similar lines. Men died not because the mine had no second opening, but because stable employees had "lost their heads at a critical moment."[1] Charges of miner carelessness were also common in the reports of state mine inspectors. Even Crystal Eastman, in her 1910 exposé of the Pittsburgh mining industry, concluded that victims were responsible for 35 percent of fatal mine accidents.[2]

It was one thing to subscribe to the theory that miner carelessness was responsible for most mine accidents. It was quite another to conclude that accidents could not be prevented. That conclusion was drawn often enough, however, to elicit this comment from Eastman: "Those emphatic, reiterated assertions, those tales of recklessness often repeated, have grown into a solid, inert mass of opinion among business and professional men in the community, a heap of unreasoned conviction."[3] The official opinion of the Industrial Workers of the World (IWW) was similar:

> It is customary for the miner [*sic*] owners and officials to blame all accidents on the "carelessness," "neglect," or the "ignorance" of the mine workers, but if such subterfuge is too raw to pass muster before public opinion, they blame it on their "God" or "Providence." However, experience has proven that where strict rules have been adopted and enforced, "carelessness, neglect, and ignorance" disappear as if by magic and the death-dealing God or Providence keeps out of the mine. This only goes to show that if the rules and regulations for coal mining were made by honest men without regard for immediate profit, and if they were rigidly enforced, mine accidents could be brought down to the same level as in the most favorable occupations.[4]

The IWW was only one participant in a small but significant reaction against miner carelessness as an operative concept. Conceding that in some sense, every accident was the result of negligence or oversight on the part of the man injured or his fellow workers, *Coal Age* acknowledged the possibility that certain precautions taken by mining officials would make some accidents unnecessary. Since under workmen's compensation the operator had to pay the bill anyway, the journal concluded that it was useless to blame others. Edward O'Toole, general superintendent of the United States Coal and Coke Company, Pittsburgh, expressed the viewpoint of the more enlightened operators:

> I have no doubt but that you have heard excuses I have often heard —"It was the man's own fault that he got killed, for he was told to set his timber, and he knew there was danger, etc."—I believe the time has now come to quit blaming the dead man; and I hope the Bureau of Mines is waking up to the fact, that it is not only the duty of the

officials to point out and tell the men of the dangers, but it is their duty to see that the dangers when recognized and pointed out are removed before they permit a man to work the place.[5]

At issue was not so much the fact that miners were careless, but the source of that carelessness and what to do about it. Miners and the *United Mine Workers Journal*, if they acknowledged the issue at all, argued that miner carelessness was a direct result of the policies of callous coal operators, who thoughtlessly employed inexperienced, ignorant labor. The *Journal* scored an anti-union mine in Webster County, Kentucky, for importing Negro agricultural workers from the Deep South, "where living is cheap, wages correspondingly low, organization unknown. These men are sent into the to them unknown dangers of the mines, without the experience which would make for their safety."[6] Mine inspectors consistently viewed carelessness as the result of the lack of English-language capability, though this analysis ignored the large number and proportion of accidents in which English-speaking miners were involved. James Roderick first raised this issue in his annual report for 1900, and he continued to be the outstanding proponent of the view that safety and language were closely related.[7] Foreign miners also were charged with illiteracy and low intelligence. The present discipline problem in the mines, said *Mines and Minerals*, "is due to the large number of aliens. . . . These men are for the most part illiterate, and of a lower standard of intelligence than their predecessors of some years ago." *Coal* agreed, but rejected *Mines and Minerals'* solution of more discipline in favor of waiting for a gradual increase in intelligence among miners. "This must be a process along evolutionary lines; it cannot be effected by revolutionary [i.e., legislative] methods." "The Slav," wrote *Coal and Coke Operator*, "is a man who bears on his mental and physical make-up the effect of centuries of governmental and social oppression. . . . Being put to work early in life, the children are stunted in mentality and physique, as a rule."[8] The problem of miner drunkenness also was raised, though less often.[9]

Carelessness was not always associated with drunkenness, lingual or intellectual disabilities, or illiteracy. Contemporaries recognized the unique characteristics of the occupation and concluded that these characteristics were related to safety attitudes. Although supervision at the working face was increased by statute in the Progressive period, even in 1920 most states required only one visit by a mine official to each working face per day. Thus, particularly in large mines, the miner retained a large degree of independence and freedom from control. Unlike factory workers, most miners worked alone or in pairs, began and finished work on their own schedules, and while at work labored as much or as little as they desired and at their own pace. Except for delays in the work (which, according to one union leader, were themselves responsible for fostering miner independence from the boss), miners were isolated even from their co-workers. Historically independent, miners grew jealous of their independence and hostile to supervision and advice. "Anything that makes me mad," said one unemployed miner, "is a boss trying to bulldoze me around—that's the worst thing I hate in the damn mine—and they don't get away with it either. I quit 'em and get the hell out of there. I know how to mine coal, the boss can't tell me nothin'."[10] According to one Illinois inspector, the miners disliked face bosses, "classing them as dictators."[11] Miners insisted that supervisory personnel had no business telling them where to set their timbers and, because they commonly paid for the explosives they used, miners questioned the boss's right to dictate where they placed their shots. Inspectors pictured the miner as fearless, proud, willfully neglectful, and hardened to danger, and they were often critical of the safety attitudes of older, more experienced miners who seemed to feel that their years in the mines had provided all necessary knowledge and justified their independence.[12] "I am frequently told by miners," said inspector J. E. Sheridan from his experience in the New Mexico territory, "when I instruct them to make their places safe, that if they are willing to take the risk, they don't see where it concerns me, if they are in jeopardy, that they should be allowed to attend to their own business."[13] This traditional independence caused one businessman to rate ex-miners as poor prospects for subsequent employment. "The ex-miner," he noted, "resents all suggestions as to his working methods, resents all effort to compel continuous application, and assumes in general, a hostile attitude to all supervision."[14] One theory associates this hostility less with independence than a form of group solidarity derived from the miner habit of congregating around piles of gob (slack, waste) when work stopped. "It's the gob pile oration," commented one miner, "that makes the miners less submissive than other workers."[15]

Other features of the coal business led miners to disregard danger. The method of paying the miner—by the ton of coal mined—encouraged him to get out as much coal as he could. To get out a good day's production, recalled former miner John Brophy, "an experienced miner would often work calmly on under conditions that would terrify a novice."[16] The irregular nature of employment, making each hour in the mine more precious, encouraged the miner to work as fast as possible. And a miner did not just enter the mine and begin to haul out coal. He was often required to make elaborate preparations of his working place (draining it of water, setting timbers, etc.) before he could begin to make his cuts and plant his shots. Because the nonunion miner received no direct compensation whatsoever for this "dead work," he was tempted to do it quickly and without sufficient care. Companies attempting to cut the death rate from falls of roof and coal would increase supervision rather than adjust the miner's pay to cover dead work. These circumstances contributed to the miner's reputation as an impatient, hasty, and therefore careless worker, but few observers realized that the miner was a victim of coal-industry economics. Two advisers to the Bureau of Mines, for example, acknowledged the economic motivation for carelessness but labeled it, somewhat harshly it would appear, greed.[17]

Evaluation of the relationship between the foreign miner and safety presents particular difficulties, since contemporaries seldom approached the subject with objectivity. One scholar, an official of the 1910 census, said that as long as accidents happened, the "best interests of society will be served by permitting the least valuable members of the mining communities to be the victims." To F. J. Warne, Slavs, unmarried and childless, were the least valuable members.[18] Lyman Beecher Stowe reported mentioning the large loss of life in the mines to a corporation executive who replied: "Yes; but, after all, it's not so serious, because most of the men killed are ignorant foreigners who can be easily replaced."[19] Others believed that discipline, obedience, and respect for mining law were Anglo-Saxon characteristics, innate or the product of American institutions.[20]

In the first two decades of the century, the new immigrants from southern and eastern Europe moved into the nation's coal mines in response to the demands for mine labor caused by the sharply rising demand for coal. The years of heaviest immigration, 1905 through 1914, paralleled the first realization that an industrial accident problem of monumental proportions existed. The worst year in the history of industrial accidents, 1907, was presaged by two years of very heavy immigration.[21] Did the new immigrants in any sense cause the safety problems? Were they more inexperienced than the old immigrants? Were they less literate? Were they less able to speak English? Were they, in short, more likely than their predecessors to exacerbate the mine-accident problem?

Unlike the English, Irish, and Germans, who dominated immigration rolls until the 1880s, few of the new immigrants had been employed as miners in their country of origin. The new ethnic groups moving into the mining industry—Slavs, Northern and Southern Italians, Croatians, and others—were agricultural peoples, without mining experience. Some 7.7 percent of the Southern Italians engaged as mine workers in America had previous experience in the mines; 13.7 percent of the Northern Italians; 9.8 percent of the Poles; 3.6 percent of the Croatians; 10.9 percent of the Magyars; and 10.7 percent of the Slovaks; in contrast, 88.2 percent of the Scottish and 55 percent of the German mine workers had mining experience abroad, but like the Irish and English, their influence in the mines was declining rapidly.[22]

That lack of mining experience was directly related to injury and death was a common assumption and one supported by logic. When bureau director Manning said a mine was "all wondrously strange and intricate"[23] to the new miner, he was understating the case. Mining was no simple occupation to be learned in a few days. Timbering, setting of shots, use of the pick and powder, and the judgments that went with these skills were learned only gradually. On the basis of common sense, the *United Mine Workers Journal* seems justified in concluding that "the peril of the mine lies in the immigrant ship."[24] Contemporaries, however, went beyond logic and common sense and attempted to prove the point using evidence that was, by itself, manifestly irrelevant. West Virginia, for example, published information correlating years of experience in the mine with absolute numbers of fatalities, thereby implying a relationship between inexperience (if not the new immigration) and high death rates. But since the number of employees at each level of experience was not provided, the state's data proved nothing. This statistical effort tells us considerably more about the values and assumptions of the West Virginia Department of Mines than it does about the causes of mine accidents.[25]

Of the various language abilities, reading English was most valuable to the mine worker, since that skill would enable him to comprehend posted mine rules and regulations, instruction manuals, and the *Miners' Circulars* published by the Bureau of Mines. Data collected by the Immigration Commission in 1909 and 1910 indicate that 96.9 percent of native-born white bituminous coal mine workers of native-born fathers could read English; only 75.3 percent of native-born

Negroes of native-born fathers had similar capability. English, Irish, Scottish, and Welsh miners were also likely to be able to read English. There are, unfortunately, no data indicating the English-language capability of any other mining group. To some extent, literacy in one's native language would substitute for English, since mining materials of the type mentioned were occasionally printed in foreign languages. A miner was likely to be able to read if he was born in this country (regardless of his ethnic heritage) or if he belonged to the older immigrant groups. Reading ability in native language was high among Northern Italians (87.8 percent), Magyars (89.2 percent), and Slovaks (82.4 percent) and low among Southern Italians (65.7 percent) and Croatians (66.9 percent).[26]

While there are no statistics for English-reading capability among most foreign-born miner groupings, the Immigration Commission did collect data for English-speaking capability. This, too, was a valuable asset for the miner, for it allowed the novice to converse with other English-speaking miners, many of whom had experience in the mines, with various mine officials, and with the mine inspector. Such contacts would, presumably, accelerate the learning process and encourage adherence to mining laws. Table 7 reveals that of the major reporting groups, some—Magyars, Poles, and Slovaks—acquired English-speaking capabilities more slowly than others.[27] More important, it is apparent that these groups maintained their native language for some time. Of the reporting non-English-speaking miners who had been in this country for more than five but less than ten years, more than 30 percent still did not speak English.

These statistics present a strong logical case for the assumption, so crucial in contemporary analyses of mine safety, that the foreign-born, new-immigrant miner was a serious obstacle to accident prevention. "It is clear," said the Immigration Commission, "that the employees of the races of southern and eastern Europe, having had little experience in mining either in this country or abroad, are particularly liable to accidents. And as the responsibility for accidents rests in most cases with the men injured, to say that they are particularly liable to accidents is in effect to say that they are responsible for a considerable proportion of all the accidents occurring in the mines."[28] The commission did not, however, provide irrefutable proof that the new immigrants were "particularly liable to accidents." The case for West Virginia was the strongest (see Table 8).[29] White Americans in that state were some 46 percent of the reporting work force yet accounted

TABLE 7

Percentage of Foreign-born Employees Who Speak English, by Ethnic Group and Years in the United States, 1910*

Ethnic group	Number reporting	Under 5 yrs. %	5–9 yrs. %	10 or more yrs., %	Total %
Bohemian and Moravian	735	43.3	72.6	91.0	72.2
Bulgarian	172	19.2	94.4	87.5	30.2
Croatian	2,394	47.2	68.1	78.9	57.8
Dutch	101	66.7	78.9	96.2	84.2
French	760	36.8	65.9	90.0	70.9
German	2,639	63.9	85.1	97.4	90.5
Greek	111	52.6	74.2	91.3	66.7
Italian, north	6,528	40.3	71.5	86.7	62.2
Italian, south	4,188	43.2	70.5	85.4	60.8
Italian (not specified)	103	40.4	77.4	85.0	60.2
Lithuanian	1,870	47.9	80.6	89.3	75.5
Magyar	4,470	34.4	65.0	78.9	51.9
Mexican	107	67.9	81.5	75.0	74.8
Montenegrin	136	29.9	64.3	100.0	36.0
Polish	7,190	30.5	60.1	78.8	52.1
Roumanian	151	43.9	66.7	100.0	47.7
Russian	1,810	41.5	73.9	87.4	60.4
Ruthenian	300	20.9	55.1	84.9	44.3
Servian	127	43.1	75.5	80.0	61.4
Slovak	11,137	35.3	63.2	80.7	58.6
Slovenian	1,864	42.2	72.1	83.3	60.9
Swedish	306	60.9	94.4	100.0	96.4
Total	48,656	38.8	68.2	85.6	61.2

*Includes only non-English-speaking groups with 100 or more males reporting. The total is for all non-English-speaking groups.

TABLE 8

Mine Fatalities in West Virginia, 1904–1908, by Ethnic Group

Ethnic group	Distribution of employees, 1908		Distribution of fatalities, 1904–1908	
	Number	%	Number	%
American (white)	23,979	46.3	551	35.5
American (black)	11,270	21.8	254	16.4
Italian	6,046	11.7	311	20.1
Polish	1,901	3.7	94	6.1
Austrian	1,013	2.0	19	1.2
Russian	851	1.6	15	1.0
Slavic	620	1.2	96	6.2
Lithuanian	506	1.0	35	2.3
English	488	.9	34	2.2
German	430	.8	24	1.5
Irish	264	.5	7	.4

for only 36 percent of the fatalities; Italians, on the other hand, were 12 percent of the reporting work force but 20 percent of the fatalities. Still, the number of fatalities is relatively small and susceptible to substantial changes through even one major disaster. Furthermore, in Pennsylvania, where the new immigrant ethnic groups formed an even larger percentage of the total work force, it is not at all clear that such groups were more liable to accidents than others (see Table 9).[30] Native white and black workers represented about 15 percent of Pennsylvania's mine labor force in 1910; yet from 1899 through 1912 they accounted for more than 18 percent of the mine fatalities. The Slavs, Italians, and Poles, on the other hand, appear to have been fatality victims in rough proportion to their percentage of the bituminous

TABLE 9

Mine Fatalities in Pennsylvania (Bituminous), 1899–1912, by Ethnic Group

Ethnic group	Distribution of employees		Distribution of fatalities	
	Number	%	Number	%
American (white)	6,448	13.1		
American (black)	913	1.9	1,216*	18.54*
Bohemian	457	.9	32	.49
Croatian	1,971	4.0	17	.26
English	1,312	2.7	212	3.23
French	339	.7	40	.61
German	1,538	3.1	211	3.22
Greek	75	.2	12	.18
Irish	663	1.3	109	1.66
Italian, north	3,379	6.9	884†	13.48†
Italian, south	2,239	4.6		
Italian, unspecified	99	.5		
Lithuanian	640	1.3	101	1.54
Magyar	3,528	7.2	96	1.46
Polish	6,025	12.3	762	11.62
Roumanian	105	.2	3	.05
Russian	1,283	2.6	225	3.43
Scottish	562	1.1	98	1.49
Servian	86	.2	1	.02
Slovak	9,998	20.3		
Slovenian	1,560	3.2	1,214‡	18.51‡
Swedish	216	.4	63	.96
Welsh	191	.4	42	.64
Austrian	247	.5	485	7.39
Belgian	153	.3	28	.43

*Includes white and black Americans
†Includes all Italians
‡Includes Slovak and Slovenian

mining population. The group that Roderick called the "English-speaking races"—American, English, Welsh, Scottish, Irish, and German—represented 24 percent of the mining population but accounted for 28 percent of the fatalities over the 1899–1912 period.

Taken together, the two sets of figures indicate the possibility—but only the possibility—that the new immigrant miner was an obstacle to progress in mining safety. At the very least, that assumption should have been subjected to rigorous testing. Rather than interpret its evidence, the Immigration Commission believed what it wanted to believe.[31]

What should have been a tentative interpretation of the accident problem became an article of faith, and the foreign miner was thrust into the center of the mine-safety movement. Reformers focused on the language problem. The Bureau of Mines, wrote Manning, "has made much progress in its safety work among the miners, but finds that its work is being greatly hindered by the large number of non-English foreign-born, who are unable to read, write, or understand the English language and in a good percentage of cases even unable to read or write their own language."[32] E. E. Bach of the Ellsworth, Pennsylvania, Collieries, said " 'English First' is the greatest asset to Safety First in both government and industry."[33] Native American and older immigrant employees, too, complained of the safety hazards posed by those who did not speak English. In response to these critiques, the safety movement after 1908 took on an educational hue. Universities in Illinois, Colorado, Kentucky, and other states offered institutes and short courses to train miners; the YMCA and the larger coal companies established night schools so that foreign miners could learn English. The Bureau of Mines used motion pictures to instruct miners who did not speak English.[34] For some, education, with its implication that the essence of the safety problem lay in miner deficiencies, served essentially as a defense against unwanted state legislation. Unless the miners were educated, said *Coal*, "a code of laws as big as the State Capital will be of no service."[35]

During the second decade of the century, this practice of blaming the victim was inherent in the Americanization movement, and education for safety and education for citizenship became virtually indistinguishable. The YMCA began to include instructions for obtaining citizenship papers in its miner education manuals; the National Safety Council assumed Americanization was essential to progress in safety. One operator claimed to see results in the Americanization program administered by his company: "For two or three years we have conducted classes, and have specialized on a class in Americanization or naturalization. Our effort there has been to cause all these men to want to be good Americans, to instill in them a desire for citizenship, and that has worked out very well. . . . We try to teach the words and danger signs and the things they need most when they get into the mine. That has done a good deal to reduce our accident ratio among the men who don't speak English."[36] Neither of the bureau directors in this decade was opposed to a blend of Americanization and education. In 1910, in response to a suggestion by John Mitchell that *Miners' Circulars* be printed in languages other than English, Holmes said: "As this would be a step in the line of good citizenship, all representatives of the Government naturally would prefer to encourage the miners to read English. . . . There is no possibility of making a good citizen out of a dead miner." Almost ten years later, supporting a federal bill "to promote the education of native illiterates," Manning emphasized that the bill would not only save miners from injury and death, but would, in addition, make them good American citizens. "My plea, therefore," he concluded in a lengthy memorandum on the foreign-born in the mining industry, "is that such a measure as the proposed Americanization bill will not only save human life and suffering in the mining industry, but will also be of great economic benefit to the stat [*sic*], the industry, and the men; it will make loyal American citizens out of men who through their ignorance remain aliens; and it will nip in the bud such movements as Bolshevism, etc."[37] Manning's exhortations received a sympathetic hearing from those who shared the bureau director's fears of industrial unrest. "Every good citizen," wrote Iowa Senator Albert Cummins, "must favor some such legislation as you are proposing. There is nothing more important than the education of these people in our efforts to secure peace, order and efficiency in the industrial world."[38] At this date the Americanization movement had the support of much of organized labor, including the United Mine Workers,[39] but while industrial safety continued to benefit from Americanization throughout the first two decades of the century, by 1920 the movement to make "good citizens" of immigrants was much less an educational device than a tool for social control.

Another solution to the problem of miner carelessness was "discipline," a word used to excess in the literature of the time and one harboring a variety of approaches to and analyses of mine safety. In a general sense, discipline meant compliance by mine workers with state mining laws and state and company regulations; it was an end product, synonymous with enforcement of the law. Beyond this generalized goal, however, discipline embodied two concepts of responsibility: that of the miner who, unsupervised most of the time, had to make responsible decisions about safety; and that of the miner who could make such decisions only if he were subject to strict supervision. The former recognized the individually responsible miner as the key element in the safety matrix; those who held this view of discipline believed that the miner must be encouraged to internalize the values of safety and act on them. This would likely be accomplished through education, but other tactics, like prosecution and dismissal, were also permissible.[40] In contrast, the supervision side of discipline presented the miner as the object upon which more basic elements—the mine officials and inspectors—would operate; here discipline was a function of management and bureaucracy rather than labor. Implementation meant more inspectors, more regular inspection, and particularly more supervision at the working face, either by mine foremen or some new breed of face boss. Europe served as a somewhat mythical model for these efforts. "In European countries," noted the bureau's Edward W. Parker in a plea for "military discipline," "operations are under strict police surveillance."[41] Interestingly, in view of the industry's emphasis on the troublesome immigrant, the foreigner was often characterized as more amenable to discipline than the native American miner. American workmen were too independent, too intolerant of restraint to be receptive to disciplinary measures; foreigners were more willing to work systematically and under close supervision. The non-English-speaking miner, wrote Illinois labor commissioner David Ross, "responds more quickly to discipline [because] a part of his home education has been to respect the orders of those in authority, and he will do without question what he is directed to do."[42]

Certification—the licensing of mine officials and miners by the government—was the miner answer to the problems of employee inexperience, carelessness, and incompetence. This effort to insure that mine employees were competent to their tasks was embodied in legislation passed as early as the 1870s. In this early legislation, however, the employer, rather than an independent board, was the judge of competency. In the 1890s Illinois, among other states, began to determine competency of mine officials—fire bosses and mine managers, for example—through a state board of examiners. Miners applauded the passage of such legislation insofar as it contributed to safer mines, but before 1900 certification was applied only to mine officials; miners were not included. Five years later, coal-producing states such as Ohio, West Virginia, Kentucky, Colorado, Oklahoma, and Kansas had no certification of any kind. Although miners continued to support examination or certification of mine officials, their major concern after 1900 was certification of their own numbers. The Special Committee on Legislation of the United Mine Workers argued in 1905 that the fellow servant defense then commonly upheld by the courts was inconsistent with selection of the labor force solely by the employer.[43]

The central goal of the miner and his union in supporting certification was increased safety. To some extent certification was seen as a substitute for the failure of inspectors to enforce and of mine bosses to heed the mining law. Under these conditions, argued the *United Mine Workers Journal*, explosions could only be prevented by an apprentice system, under which the applicant would serve a term of years under a practical miner and then take an examination. Occasionally miner arguments spoke the language of professionalization, albeit mixed with the fundamental safety argument: "No man is allowed to practice law or administer medicine unless he is declared competent by a proper tribunal. . . . Why should not the same rule apply to mining? If the law can protect people from quacks and pettifoggers why should not the mine workers be protected from the ignorance of their fellows?"[44] Operator fears to the contrary, worker control of labor markets appears to have been a minor motive in labor's support of miner certification.

John Mitchell believed fundamentally that inadequate legislation and law enforcement more than miner incompetency were the essence of the safety issue, yet it was he who took up the leadership of the competency crusade. In 1906 he presented to the UMWA National Convention, for passage by the states, a bill providing for certificates of competency for all miners and the creation in each county in each state of a miners' examining board, to consist of three miners appointed by a circuit judge. This board was to grant a certificate of competency to a miner with two years' practical experience as a coal miner and with the ability to answer at least twelve questions in the English language. In what was essentially a drive for uniform competency laws, Mitchell had the support of the national miner organization and its leaders, including Van Amberg Bittner, John Walker, and T. L. Lewis.[45] National safety organizations not controlled by the miners did not support miner certification. The Mine Inspectors' Institute, which might have been expected to favor miner certification and to press for its uniform application in the states, did not see fit to discuss the issue, an omission even more obvious when viewed against a background of MIIA advocacy of certification of all mine officials. Oddly enough, Joseph A. Holmes and the Industrial Workers of the World held similar conceptions of the appropriate place of certification: a limited examination would precede the real education process, to be carried out in a mining school.[46]

Operators were divided and ambivalent. Sympathetic to the general principle behind the legislation, they typically feared, and with some justification, that certification, if administered by miners as Mitchell and others suggested, would give labor control of the supply of workers.[47] Registering these apprehensions, *Coal* warned that certification laws would confer great political and industrial power on the mine workers' unions: "There will be a howl when certificated miners cannot be obtained to work [the mines]."[48] To operator Justus Collins, passage of a certification law would force employment of certified miners and lead directly to union dominance in West Virginia. In Pennsylvania, operators successfully opposed a certification law for the bituminous region which would have required at least two years' mining experience in order to qualify for mining coal at the face. Miner-controlled examination boards were, in fact, in an ideal position to prevent the importation of scabs during a strike in a unionized district, to curtail the labor supply in order to provide steadier employment for established miners and raise payment rates, and even to unionize previously unorganized areas. The anthracite miner certification law was used during the 1900 and 1902 strikes to deter importation of scab labor. A number of trade journals, nonetheless, recognized the legitimate benefits to be derived from miner certification. It would save lives and property, protect the miner from cheap foreign labor, and, perhaps most important, relieve operators from responsibility for hiring incompetent miners. But the divisions among operators militated against an effective certification campaign, and of the major coal-mining states, only Illinois and Indiana had effective legislation. Ohio and Montana maintained examination in word only, since the mine boss functioned as the examination board. As of 1915, West Virginia, the Pennsylvania bituminous region, and other, less important coal-mining areas had no provisions for miner examination and certification.[49] Even where enacted, certification did not immediately fulfill miner expectations. In Illinois positions on the examination boards became political plums, and there and in the Pennsylvania anthracite region, certificates of competency were sold and granted in exchange for votes. Insofar as certification requirements were supposed to protect the miner from non-English-speaking foreign labor, they largely failed; immigrants were known to have passed the language requirements with interpreters or through friends.[50]

The activity of the United Mine Workers in behalf of certification was only one of numerous ways in which the union promoted safety in the coal mines. Nationally the UMWA was influential in creating the Bureau of Mines in 1910, in securing its appropriations, and in obtaining the rescue and experiment stations in 1915. It accomplished these goals with only the most perfunctory aid from the parent organization, the American Federation of Labor, whose priorities emphasized eight-hour legislation, amendments to the Sherman Antitrust Act, and opposition to the injunction and convict labor. The cooperative safety activities of the mine workers, particularly in conjunction with the Bureau of Mines, were substantial, and after 1905 the union's mouthpiece, the *United Mine Workers Journal*, devoted a good deal of space —both in its regular columns and on its editorial page—to the political and technical aspects of coal-mining safety and to the more general problem of industrial safety.[51] In the states, the miners' union urged selective support of political candidates, pressed for enforcement of existing safety laws, and was responsible for the formulation and passage of much new legislation. The tools available to the union as collective bargaining agent for the miners—the contract and the strike —were also used to pursue safety objectives, but neither was fully exploited and each was occasionally used to achieve objectives inimical to safety. And with the exception of the years from 1905 through 1910, the union understandably emphasized organization and membership and made safety a minor and peripheral issue.

The United Mine Workers inherited these priorities from its predecessors, for the goals of membership and recognition dominated the early organizational efforts in the bituminous fields, where in every case safety was a secondary objective. The first major bituminous union, the American Miners' Association, was organized in West Belleville, Illinois, in 1861; throughout its seven-year existence, the association was absorbed with questions of wages and methods of weighing coal. It nonetheless laid the groundwork for safety legislation in the

1870s through the support of political candidates who favored the miners' legislative demands. Local unions persisted from the demise of the American Miners' Association in 1868 until 1873, the date of the formation of the Miners' National Association from Illinois, Ohio, and Pennsylvania local groups. Its constitution, like those of other miner unions, listed safety and health as organizational objectives, but the union was too preoccupied with maintenance of the organization through the depression to pursue seriously the safety issue. Local organizations again held the field from 1876 to 1885, when leading unionists from Illinois, Indiana, Ohio, Pennsylvania, West Virginia, Iowa, and Kansas organized the National Federation of Miners and Mine Laborers. Although concerned primarily with ventures in national collective bargaining (setting wage rates) and conflict with Knights of Labor National Trades Assembly 135, the union did recommend political action to obtain stronger safety legislation, particularly in the area of child labor. Resolution of the complex labor conflict began in 1888, when the National Federation and a portion of the Knights of Labor National Trades Assembly 135 merged as the National Progressive Union, and was completed in 1890, when the remainder of Trades Association 135 joined the National Progressive Union to form the United Mine Workers of America.[52]

To the UMWA, faced with roughly the same problems and obstacles as the earlier unions—maintenance of membership, recognition of the union, opposition of employers, low wages, and countless unfair labor conditions, safety was a peripheral concern. The issue took a back seat to attempts to abolish company stores and the screen coal system and to obtain checkweighmen, weekly paychecks, and compensation for dead work. The establishment in 1898 of the Joint Conference in much of the central competitive field did not solve the union's problems, for membership in 1898 was only about 32,000, up from 9,700 the year before. The end of overt opposition to the union in much of that field, however, enabled the UMWA to increase its membership to more than 250,000 by 1904. Although the union continued to be faced with troublesome unorganized districts (West Virginia and Alabama, for example), the organizing successes of the previous years, coupled with the turn of safety conditions from bad to worse and the strong leadership of John Mitchell, enabled the mine workers to devote a proportionately greater share of their energies to safety, producing something approximating a miner campaign for coal-mining safety in the years from 1905 to 1911.[53] At the 1908 annual convention of miners in Pennsylvania's District 2, Pittsburgh region organizer Edward McKay captured the new interest in safety as he asked the delegates to resist the introduction of dangerous machinery and to reevaluate the importance of working conditions. "We have more to think of in our conventions," he said, "than the mere settlement of wage scales."[54] Across the coalfields, union miners responded to similar pleas with major political efforts, particularly in the states. The upper echelon of the union hierarchy was the slowest to move on the safety issue. At the March 1909 meeting of the union's National Executive Board, for example, the rash of recent disasters was not on the agenda, members preferring to discuss possible amendments to the Sherman Act, an Alabama strike, and appeals of union locals for financial aid. At the peak of national interest in coal-mining safety, one member of the executive board admitted the board had paid little attention to participation on a committee designed to prepare compensation and safety legislation; another cautioned the organization not to "remain dormant as we have in the past."[55] T. L. Lewis's 1910 presidential address couched the same ambivalence. Although Lewis mentioned the Bureau of Mines and the Cherry disaster, organizational struggles dominated his presentation.[56]

The end of the UMWA flirtation with safety was signaled by new organization and membership difficulties. Between 1905 and 1911, union membership actually decreased by some 8,000. The situation became critical in 1912 when operators in the only organized region of West Virginia refused to renew the contract on the agreed basis. Union attention was drawn to this and other unorganized areas. A series of important strikes and what president John P. White called "well organized conspiracies within our own ranks to destroy the organization" in Districts 2 (central Pennsylvania), 5 (Pittsburgh), and 10 (part of Ohio) absorbed union energies in 1912.[57] Although the union did not completely eschew legislative activity, safety was eclipsed as the mine workers worried about convict labor, Taylorism, the injunction, and long hours. The four years after 1911 were difficult ones for the UMWA, with unemployment (particularly in the fall and winter of 1914) a threat to union stability. "Because of the unfavorable industrial conditions which prevailed during the past two years," said White at the 1916 convention, "the strength and efficiency of our organization have been severely tested and the ingenuity and ability of those in charge of the affairs of the organization have been tried as never before." Mother Jones echoed White. "This administration," she said, referring to struggles in West Virginia and Colorado, "has had more fights on its hands, more to go up against than any administration you have had since you were organized."[58] Under these pressures, and faced with fewer and less severe disasters than in the preceding ten years, interest in safety, at least in the national conventions, declined considerably after 1912. Safety received only a few stylized comments from the president in his annual report and an address (generally routine) by the director of the Bureau of Mines. The war, while stimulating organization, only added to the issues considered more important than safety, and in 1918, for the first time since 1900, the president's report to the convention contained nothing about safety, the subject having been relegated to the union vice president. The *United Mine Workers Journal* muzzled its powerful safety voice and responded instead to the urgent call of David Lloyd George, then British minister of munitions: "Coal means life for us and death for our foes. Steam means coal, rifles mean coal, shells are filled with coal, the explosive inside them is coal, and coal carries them right onto the battleship to help our men. Coal is everything to us and we want more of it to win the victory."[59] The union demand for nationalization of the coal mines in these years was based not on considerations of safety but on a desire to curb the waste of resources which the mine workers saw as a concomitant of the competitive system.[60]

Following a 1938 mine-safety conference of operators, miners, and public officials, E. A. Holbrook, then dean of the University of Pittsburgh School of Engineering and School of Mines, wrote to John B. Andrews of the American Association for Labor Legislation. The conference, Holbrook said, "brought out what I have long felt—that neither the mine operators nor the mine workers as organized bodies, know much about mine safety. After all, one cannot bargain collectively with mine safety."[61] Whether or not one *could* bargain collectively over mine safety, it was never done on any grand scale. Certainly the potential was there. Oklahoma's chief inspector, Peter Hanraty, suggested that safety was an appropriate subject for miners' meetings and for their contract negotiations with operators. From 1898, however, when collective bargaining was firmly established through most of the central competitive field, through 1930, the Interstate Joint Agreements were silent on safety issues. While some of the interstate contract provisions may have had indirect effects on mine safety (by stimulating production, for example), neither the original contract of January 28, 1898, nor its successors contained any mention of safety.

In presenting their bargaining position to the 1912 Joint Conference, the miners argued that machinery, electricity, and the use of oil and gas as fuel had increased the dangers in the mines. But instead of using the contract to decrease the dangers, the miners used the dangers to justify a series of demands with a heavy economic content.[62]

Most contract provisions with relevance for safety were inserted at the district (or state) conference. Typically the district contract defined some of the duties and rights of the operators. Operators were often required to keep on hand sufficient first-aid supplies, to furnish powder for blasting and props for roof timbering, to construct air courses which would assure proper or legal ventilation at the working face, and to provide shields for machinery. Occasionally a specific provision required the operator to go beyond state laws, as in Wyoming where the 1909 contract called for twice-weekly sprinkling of dusty roads and entries. Operator rights included the right to designate the kind and quality of explosives. In Illinois miners argued for a clause in the contract which would make the mine examiner (fire boss) a representative of the miners rather than of the operator. They succeeded in this and in incorporating into the 1904 contract a provision for discharge or transfer of an examiner found, after adjudication, to be incompetent.[63] For his part, the miner was bound, under most district contracts, to timber properly and otherwise care for his working place and to shoot the coal "correctly." The 1906 Illinois agreement was most explicit: "The United Mine Workers of Illinois shall continue to assume all responsibility heretofore resting upon them for the care of the working places and the proper character and placing of the blasting shots."[64] Some contracts required the miner to modify mining machines in order to make them safe for transporting. And they almost invariably included actions which the operator could take if the miner, through failure to carry out his contractual obligations, posed a safety problem. The Kanawha District contract, for example, provided that "any miner will be subjected to discipline who, from ignorance, carelessness, or any other cause, fails to properly mine, shoot and load coal."[65] Failure to timber the working place properly was almost always an offense for which the miner could be discharged. Illinois miners objected to a similar but tougher clause in the District 12 contract which, they said, allowed the employer to fire an employee for failure to live up to the state mining law, but which "leaves the other parties unbound to carry out the laws of the state except as good citizens."[66]

Shot-firing, a controversial issue, was the subject of numerous district contract provisions. One contract required that shot-firers' wages be paid by the operators, another that they be paid by the miners. The 1903 Indiana bituminous contract established a commission of one miner, one operator, and one public representative to determine if shot-firers were necessary in individual mines and relieved the Indiana operators of legal responsibility for shot-firers at work in their mines. In Kansas the contract provided for shooting off the solid, a mining method prohibited by state law. Disputes over shot-firing were a regular feature of negotiations in Illinois, where in 1904 the miners came to the bargaining table asking that shot-firers be employed in mines which were unusually gaseous, not properly ventilated, or where blasting was done on the solid. The operators responded predictably. "If a man is not good enough to fire his own shots," said Richard Newsame, "he is not good enough to go into a coal mine. . . . all the operators are asking is that their coal shall be mined in a workmanlike manner, and then we will have no explosions from shots."[67] These simplistic arguments indicate that the shot-firing question, while ostensibly one of safety, was a matter of controversy in Illinois and other districts primarily because of the expense involved in hiring personnel to fire the shots.

To adjudicate grievances and disputes arising out of the contract provisions, the contracts clarified existing arrangements and established new procedures and institutions. Contracts typically defined the province of the pit committee, an elected group of miner representatives which negotiated with the mine foreman or pit boss at the second stage of the grievance procedure and which, according to one operator, "has always contended that the miner was right and the pit boss was wrong."[68] But if the contract institutionalized the pit committee, it also circumscribed its functions. The 1904 Illinois agreement, for example, prohibited the pit committee from traveling around the mine unless called to do so by the pit boss (an operator representative) or a miner. Miners argued that since they were required to sign the reports of the mine examiner, they should be entitled to accompany him on his rounds. Another agreement prohibited the pit committee from taking up grievances during working hours and insisted that "the Mine Committee shall have no other authority whatever" outside of the adjustment of disputes. "It is understood," read this agreement, "that no Mine Committee or employee has the right under this agreement to stop work to adjust any grievance or call a strike at any mine under any circumstances whatever."[69] The object of these restrictions was to prevent the committee members from assuming management functions and from carrying on union activities (particularly from calling a strike) while in the mine. Although evidence on the performance of this informal mechanism is limited, apparently the best of the pit committees actively worked to correct unsafe conditions.[70]

Analyses of the disputes under contract in all formal stages of the adjudication process in the anthracite, Illinois, and southwestern coalfields reveal that miners seldom used the formal grievance mechanism to complain of—or to attempt to remedy—conditions within the mine. The grievance system was most often used by miners attempting to obtain compensation for time lost waiting for mine cars and mine cages to transport them into, out of, or around the mine. Miner discharge was another frequent subject of adjudication, though within this broad category lay some cases related directly and indirectly to safety. Appeal procedures were used by miners discharged for violations of safety provisions in the contract, and by miners discharged (according to the miners) for complaining about unsafe conditions. Miners were more likely to employ the grievance mechanism to secure the reinstatement of workers dismissed for safety violations than to protest an unsafe condition.[71]

No matter how elaborate, the contract grievance mechanism could not successfully deal with every dispute arising under it. When established procedures broke down, the result often was a strike or walkout. Before and after the firm establishment of collective bargaining in 1898, the causes of strikes reflected the fundamentally economic values of the union and its membership. Wage reductions were the central issue in an 1897 strike in the central competitive field which involved 150,000 men in five states and resulted in the calling of the Interstate Joint Conference. Asked by the 1900 Industrial Commission for his opinion of the causes of strikes, W. C. Pearce, secretary-treasurer of the UMWA, said: "The principal causes which lead to strikes, of a national character at least, are either to prevent a reduction of wages or demands for an increased wage." Pearce went on to suggest that safe working conditions—particularly adequate supplies of air and timber—were important to the miner but that conditions in this area had improved so much that "strikes from these causes very seldom occur now."[72]

Safety often appeared as an auxiliary issue in a strike carried on for more basic reasons. Such was the case in the Anthracite Strike of 1902, for here hazardous working conditions appeared not as one of the four basic miner demands, but as justification for the miner call for a 20 percent wage increase. "The rate of wages in the anthracite coal fields," claimed the miners, "is insufficient to compensate the mine workers in view of the dangerous character of the occupation."[73] One of the first anthracite unions, the Workingmen's Benevolent Association, used precisely the same argument in 1868. In a 1910–1911 strike in Westmoreland County, Pennsylvania, safety was at most an underlying grievance of the miners. Major demands were the familiar, largely economic ones—wages, checkweighmen, eight-hour day, recognition. Similar conditions forced the miners to strike in Colorado in 1903 and 1913–1914, although in each of those strikes safety was an issue and in 1903 the miner demand for an adequate supply of pure air constituted, officially at least, one of the five major grievances.[74]

Miners occasionally struck for safety reasons alone. Ohio miners walked out of a mine because they thought the wall between their mine and the adjacent one, which was flooded, was unsafe; in Denning, Arkansas, workers were out for nine months in a successful attempt to get the Western Coal and Mining Company to employ shot-firers; insufficient ventilation caused the miners and the shot-firers in an Indiana mine to go out on strike in 1908; and on several occasions Illinois miners refused to work when they discovered unsafe cages or gaseous mine air.[75] The union hierarchy carefully maintained the right to strike or suspend operations where life or health was endangered. "The trade unions," wrote John Mitchell in 1903, "must continue by agitation and education, by appeals to legislatures, and, if necessary, by strikes, to enable good and compel bad employers to do everything within their power to lengthen the life and maintain the health of their workers."[76] In 1910 Edwin Perry, secretary of the UMWA National Executive Board, stated his belief that the grievance machinery was preferable to the strike unless the health or lives of the miners were at stake. Usually this right went uncontested, but stoppages provoked some operators to protest. At the Interstate Joint Conference in 1908, operator John H. Jones of Pennsylvania posed the problem in these terms: "If, on account of some accident over which the management of the mines has no control, some miner without consulting the mine management, without taking the question up with the management at all, and asking for another place to work in the mines, rises up and says to the others, 'Let's go home'; that is not unionism, that is radicalism; and the man that does that is an enemy to union."[77] One can appreciate Jones's anxieties without acknowledging the problem as a serious one. Safety walkouts were always brief, seldom lasting more than a few days, and they were hardly a regular feature of labor relations in the coalfields. In the major strikes, safety was more a means than an end—a device that miners used to elicit public sympathy for their cause and to rationalize and secure the economic demands which, with recognition itself, were the essence of coal unionism in these years.

It is hardly necessary to state that miners would not, and did not, strike purposely to perpetuate or to initiate unsafe conditions. Nonetheless, miner strikes initiated in ignorance or for reasons of economics sometimes had the effect of producing such conditions. Strikes to secure payment by run-of-mine were common and, if successful, increased mine hazards. Ignorance was the major stimulus behind a 1916 walkout of 20,000 Linton, Indiana, miners to protest the introduction of Edison electric safety lamps. The miners claimed that their old open lamps burned out small accumulations of gas and made the mines safer; in fact the gas in Indiana mines was at that time approaching a level dangerous for open lamps. The problem was sufficiently serious to elicit considerable attention from the Bureau of Mines.[78]

The revolt of Pittsburgh miners against certain permissible explosives is illustrative of the influence of economics on union policy. Although permissible explosives were first used in the United States in 1901, their use was not actively promoted by any organization until 1908, when the Geological Survey began testing explosives and publishing a list of permissibles. Government experts were excited about the possibility of reducing accidents. Clarence Hall, in charge of explosives testing for the Bureau of Mines, was convinced that the substitution of permissibles for black blasting powder would result in a considerable improvement in underground working conditions. By 1915 increased use of permissibles had reduced the percentage of deaths caused by mine explosions.[79] Opposition to permissible explosives was not unusual among miners, who in 1910 and 1911 besieged UMWA national conventions with resolutions opposing their forced use. In 1909 Ohio miners contested a bill in the state legislature which would have prohibited the use of black powder, their objections based on the projected cost of the new explosives. Trouble began in District 5 when the Pennsylvania chief inspector ordered the use of safety explosives as tested by the United States government. Almost at once, strikes broke out in areas where particular explosives—Masurite and Carbonite—had been selected by the operators. The miners argued, probably accurately, that several of the explosives rated permissible shattered the coal to a greater extent than did black powder or some other permissibles. District 5 president Francis Feehan calculated the anticipated earnings reduction at five to ten cents per ton of coal mined. Among the other reasons for the walkouts were miner claims that the explosives produced a nauseous smoke and the generally higher purchase price of the permissibles.[80] The solutions proposed by the miners were not unreasonable. Some locals demanded that miners be paid for the excess slack produced, others that entire mines be placed on the run-of-mine basis, still others that general tonnage rates be raised to make up for the difference. There were no demands for a return to black powder. District officers, however, in spite of substantial evidence to the contrary, denied that these or any other permissible explosives could contribute to safer mining conditions, claimed that coal dust was not an explosive agent, and refused to acquiesce in an investigation of the problem by a committee of the National Executive Board. The board, in fact, heard a report from three of its members who claimed that the "grievances would have been amicably adjusted had not President Feehan broke off negotiations."[81] "President Feehan," noted the *United Mine Workers Journal* in its repudiation of the actions of District 5, "does not reflect the intelligence of the craft and sets himself in opposition to facts that have been demonstrated time and time again."[82] Pittsburgh region operators relished this opportunity to cast themselves in a leadership role in the safety movement and the miners as hypocrites. "For years," said the *Coal Trade Bulletin*, "the miners have made complaints long and loud against the danger of the mines, urging on the legislature, the state authorities and the mining department the necessity of lessening the danger in the pits. Now, when these authorities find a way to decrease the danger, and attempt to put it in force, the first ones to raise a 'howl' are these selfsame miners. And this in the face of a demonstration—practical in that it was made in a mine under working conditions—that the permissible explosives produced more lump coal at a less cost and with a smaller quantity of explosives than could be accomplished if black powder had been used."[83] As distorted a picture of the safety movement as this attack presented, it contained an important element of truth; the Pittsburgh episode had revealed extraordinary ignorance of elementary coal-mining safety on the part of district officials and clearly demonstrated the overriding importance of mining rates to union membership. Illinois operators failed to solve the mine-run problem but avoided a similar confrontation by agreeing in the contract to furnish permissibles at the same relative cost as black powder.[84]

As in the Pittsburgh district, miners in most parts of the nation strongly supported the mine-run system of payment. They regularly demanded payment by mine-run in their encounters with operators at the interstate and district levels, and their demands were met through the contract and state legislation. A number of areas came under mine-run through the 1898 Interstate Joint Agreement, and by 1910 Illinois, Indiana, Kansas, Arkansas, and the Kanawha field of West Virginia were under some form of mine-run payment system. Ohio adopted mine-run in 1913 and parts of Tennessee and Kentucky had rejected the screen-coal system by 1920. Mine-run payment was a high priority miner demand for several reasons. Miners suspected that under the screen-coal system some operators employed fraudulent screens and denied workers their legitimate wages. Aside from fraud, miners knew that the screening method of payment provided the operators with a marketable slack, free of charge. In addition, the spread of permissible explosives after 1908 made it increasingly difficult for miners to shoot lump coal.

The rapid transition to mine-run had significant consequences for safety. Under the screen-coal system miners were encouraged to produce large lumps of coal rather than slack; as a result, they took care to undercut and otherwise prepare the coal before placing their shots. Such preparations allowed the miner to shoot the coal with relatively small quantities of powder. With miners paid for all the coal they produced—lump or slack—the normal restraints on the miner were removed, and time-consuming undercutting of the coal was often bypassed in favor of shooting without any undercutting, a process known as shooting off the solid or solid-shooting. The pick, formerly the stock in trade of the miner, was partially replaced by larger charges of explosives. Although most systems of payment under mine-run called for higher rates for lump than for slack, miners often chose to maximize their earnings by blasting huge quantities of slack rather than smaller quantities of lump coal.[85]

To this explosion-conscious generation, the crux of the mine-safety problem was the relationship between mine-run, solid-shooting, and blown-out shots. Operators tended to tie solid-shooting directly to mine-run. Indiana's P. H. Penna, a UMWA official turned operator, said in 1906 that "the enormous loss of life is the legitimate outcome of the mine-run system." "We have never heard tell in our state," he asserted, "of blown-out shots and windy shots, of mines being blown up and men killed until the mine-run system became operative."[86] Like other operators, Penna suggested remedying the situation through a return to the screen-coal basis, a solution also advocated by the American Mining Congress. Other solutions that appeared feasible to individual operators included prohibition, by individual companies, of shooting off the solid; increased supervision of miners at the working face; and criminal penalties for solid-shooting.[87]

In the face of operator objections, miners nationwide came to the defense of the mine-run system. Fully aware of the dangers of solid-shooting, they rejected the operator assumption that the shooting method was inalterably tied to the payment method. The answer, they said, was not to return to a corrupt (at the very least, corruptible) screen-coal payment system, but to employ professionals to fire shots in the mine. Shot-firing legislation became popular following the May 1, 1900, disaster at the Winter Quarters mine in Utah, and by 1906 was in effect in Missouri, Iowa, Kansas, Illinois, and Indiana. Mitchell of the miners' union expressed the hope that shot-firing legislation would become widespread; it is, he said, "the only measure" by which loss of life due to explosions can be "reduced materially."[88] In Indiana miners at one point resolved to sign no contract with the operators until "a well drilled force of shot firers shall be employed."[89] In Illinois the struggle over shot-firing was particularly acute. Operators called the proposed shot-firing law of 1905 "vicious" and claimed the costs of employing shot-firers would price Illinois coal out of national markets. It was not long before the Illinois shot-firing controversy degenerated, like so many similar events, into petty squabbles involving a few cents per ton of coal mined. Essentially, operators wanted the system which would be least likely to increase their cost of production; the miners wanted the one which would be most likely to increase their wages. In the process operators ignored the obvious failures of the screen-coal system, and miners paid no heed to the destructive characteristics of the mine-run system and refused to acknowledge the serious hazards faced by professional shot-firers.[90]

If the contract was not used as effectively as it might have been, if through ignorance or economic need miners and union leaders failed to confront the safety issue head-on with the strike, and occasionally rationalized an inaccurate notion of mine safety, still the sum total of union actions was unquestionably in favor of safer coal mines. Organized miners had their representatives in state and national capitals, on pit committees, and at every other stage of the grievance procedure; their unity gave them a voice in politics and a voice at the mine, including some freedom from being discharged for complaining about dangerous conditions. Although the individual miner may have contributed to the mine-safety problem, the union sanctioned enforcement of the mining laws, reporting violations to mine officials and inspectors, discouraged workers from riding on loaded cars and motor trips and from disobeying mine rules, and interjected miner influence into the appointment process for inspectors. "If it was not for the trade unions," announced Ohio inspector D. H. Sullivan to the annual convention of the MIIA, "the mining laws of my state . . . would not be properly enforced."[91] That organization was a prerequisite to mine safety was an article of faith with the *United Mine Workers Journal*, whose editors seldom lost the opportunity to contrast conditions in organized and unorganized areas. When the Ziegler mine at Ziegler, Illinois, exploded, the *Journal* found it "curious that the only unorganized mine in Illinois has had two explosions in recent years, while in all the others that are organized there has been no such dire calamity happen."[92] The same charge was brought even more directly following explosions at Eccles, West Virginia, and Dawson, New Mexico. Available statistics indicate that, at the very least, the miner in a state completely organized by the union was more likely to stay alive than one mining coal in an unorganized state. In 1907 there were 2.47 men killed per one thousand employed in the thoroughly organized states; 5.07 per one thousand in the partially organized states; and 9.49 in the states without organization.[93] There was no small element of wisdom in the suggestion of the *Journal* that in more than half the cases, the verdict of the coroner's jury should be "Death was due to lack of organization."[94]

CHAPTER

5

The Operator as Victim

THE GREEDY, insensitive, vindictive operator was one of the day's most common characterizations, usually an indulgence of miners but a conception which also had wider appeal. Following the Cherry disaster, UMWA officials charged that mine officials had closed the shafts prematurely, "without regard to the lives of the miners," solely for the purpose of salvaging and protecting mine property.[1] Three hundred lives, added socialist Adolph Germer, "have been snuffed out through capitalist greed."[2] A similar picture of the coal operator emerges from "The Draped Charter," a poem authored by an Indiana miner:

Our charter is draped; there is great lamentation
There's widows and orphans, the morgue and the bier;
With fiendish glee and with grim exultation,
Death rode through the mine in a chariot of fire.
How great was the crime and how needless the slaughter—
The call of the dollar—black and bloody the deed;
It caused the heart-breaking of mother and daughter
This dividend-making, this profit and greed.[3]

Organizer Mother Jones told this story:

> I was talking with a miner's wife one day when we heard a distant thud. She ran to the door of the shack. Men were running and screaming. Other doors flung open. Women rushed out, drying their hands on their aprons.
>
> An explosion!
>
> Whose husband was killed? Whose children were fatherless?
>
> "My God, how many mules have been killed!" was the first exclamation of the superintendent.

> Dead men were brought to the surface and laid on the ground. But more men came to take their places. But mules—new mules—had to be bought. They cost the company money. But human life is cheap, far cheaper than are mules.[4]

Incidents and opinions of lesser drama but perhaps better authority also abound. A revealing exchange took place in 1909 at a legislative committee hearing in Pennsylvania to which operators and miners had been called to give their views on a major proposal for reform of the safety laws. Speaking for the miners of UMWA District 2, president Patrick Gilday recommended passage of the bill in question, though it had, he said, already been compromised by "commercialism," and he concluded his statement to the committee with some comments on miners' asthma. When he had finished, the first operator to respond said, "I have never met a miner with asthma."[5] Mixed as they are, the state inspectors' reports contain similar observations. James Blick, retiring from Pennsylvania's Seventh District after twenty years as an inspector, noted in his final report the substantial cooperation he had received from mine operators. "There are others," he continued, "both operators and officials, holding responsible positions, who always take a delight in evading the mining law, and neglecting their duties as far as possible, especially in matters relative to health and safety. Their actions would seem to indicate that their humane principles are very limited."[6]

Descriptively, this kind of evidence confirms what we already knew: that large numbers of coal operators and their hired officials did not do everything possible to make the mines safe; that they resisted enforcement of state mining laws and were consistently a conservative influence in the state legislatures; that they often spoke and acted in a manner that could reasonably be described as callous. Analytically, however, the insensitive and greedy operator is not a concept of much utility. It neither adequately clarifies *why* many operators were not sufficiently committed to safety, nor does it explain the operators who apparently had some substantial interest in safety except by implicitly endowing them with high morality and good character. The concept does not account for the substantial operator interest in establishing and funding a bureau of mines, in federal experiment and rescue stations and cars, or in the campaign for uniform state legislation, and it ignores the cooperation between operators and miners on numerous state commissions charged with developing mine-safety legislation agreeable to labor and capital.

Although no one model can encompass the behavior of thousands of individual operators, only an economic model of behavior makes sense of most of the data. According to this model, coal-mine operators were economic men, strongly influenced—even in their attitudes toward safety—by considerations of profit and loss. Within this framework, an operator might be receptive or hostile to safety, but he would make each decision on the basis of the anticipated effect it would have on his economic condition—on the physical property of his or the corporation's mine, future earnings, his own livelihood. Operator responses to a number of safety stimuli—workmen's compensation, schedule rating, the unions, public pressure—were conditioned largely by economics.

An economic model of behavior has a peculiar relevance to bituminous coal because of that industry's unique structural and performance characteristics: a low level of concentration; thousands of small firms; chronic excess capacity; minimal profits.[7] Coal-industry economics placed enormous pressures on operators to hold down costs, especially for labor but even for items, like safety, which represented a relatively small share of total costs. As a whole, coal operators were neither greedy nor motivated by any particular desire to oppress labor by depriving it of safe working conditions; in an important sense *they* were the victims, captives of an industrial structure and system which treated natural and human resources with equal disdain. Fearful of losing ground in the competitive struggle, operators opposed state legislation and administrative regulation, made much of the expense incident to safety legislation, and were reluctant to undertake experimental mine-safety work requiring heavy financial commitments.[8] From their business perspective, however, operators saw that only by transcending the states through national action would safety and competition be made compatible, and they turned to national solutions—to the Bureau of Mines and uniformity—to solve the problem of interstate competition.

An economic model illuminates different aspects of the safety attitudes of West Virginia operators A. B. Fleming and Justus Collins. A former governor of West Virginia whose coal holdings included the Clarksburg Fuel Company, the Pittsburgh and Fairmont Fuel Company, and the Southern Coal and Transportation Company, Fleming presents the classic case of the big operator turned cautious reformer. When the explosion in his Monongah mines, by reputation among the safest in the nation, touched off new enthusiasm for mine-safety reform, Fleming and his fellow operators reacted well within the norms of the economic model. Fearful that an aroused state legislature might enact costly legislation, they shifted the focus of the reform movement to the national government. Fleming was Joseph A. Holmes's strongest ally in the campaign to establish coal-mining safety securely in the Geological Survey and the Bureau of Mines. "The Fairmont Coal Company," wrote Holmes to Fleming in a handwritten note in 1909, "has shown an admirable desire to cooperate with us in everything looking to the betterment of the coal mining industry."[9] Cooperation, however, did not extend to every aspect of mine safety. Understandably chagrined at the public-relations impact of explosions (for any number of reasons, including the economic), Fleming did his best to minimize publicity in this area. Upon receiving a copy of Fleming's address to the West Virginia Board of Trade, the secretary of the West Virginia Mining Association, Neil Robinson, wrote him that he was "really quite glad that you decided to omit references to the mining fatalities which have clouded the coal industry in the past."[10] Even after Monongah, Fleming's own properties were in violation of West Virginia safety regulations. Passed February 22, 1907, Section 25 of the state mining law called for the adoption, distribution, and publication by the management of special rules for the government and operation of its mines. Fleming, in a letter to George T. Watson, manager of his Consolidation Coal Company, noted that Consolidation had not adopted the special rules required by the mining law. Some individuals and the company itself, he said, could be charged with a misdemeanor and fined or imprisoned. Fleming continued:

> I don't think there is much likelihood of anyone being indicted or fined, and I only call this to your attention incidentally; but in the event we should have some serious accident at one of these mines and if it should turn out, as it necessarily would, that no special rules had ever been adopted as required by the foregoing section, it seems to me that the company and the officers of the company, as well, might get in very serious trouble; it is very possible that the company itself would be liable in damages, without regard to what caused the accident. It is also possible that the officers of the company, in the event some one or more persons should be killed, might be held liable for manslaughter.
>
> In addition to this, there is the further possibility that the State Mine Inspectors might close up those mines until such special rules are adopted and promulgated.

Fleming, at least in this letter, did not demand of Watson that the regulations be fulfilled; and if he did imply that some attempt to satisfy the law be made, the grounds were not that lives would be saved, but rather that the company would be liable for damages, its officers liable for manslaughter, and the mine forced to discontinue production.[11] Although this incident indicates the insufficiency of the economic model and the complex nature of Fleming's safety posture, it also reveals the primacy of Fleming's interest in production and the corporation.

Justus Collins presided over a much smaller empire, but his safety views were filtered through business values to an even greater degree. Unlike Fleming, who saw the explosions of 1907 and 1908 at least partially as problems in need of scientific solutions, Collins viewed them in the context of the attempts of the United Mine Workers to unionize the state. The explosions, he concluded, were "caused intentionally. Any man could slip in three or four hundred pounds of high percentage dynamite and by applying a long fyse [*sic*] to it could set it off and be out of danger himself before the explosion occurred." Once the union had caused the explosion, Collins continued in a letter to private detective T. L. Felts, asking him to investigate the problem, the mine workers could go to the West Virginia legislature and request "drastic legislation against the coal operators and mine owners, and the union with all their following, and the demagogues throughout the State, would like to see all kinds of legislation enacted which will practically put the operators out of business."[12] Collins wrote in a similar vein to owner Isaac T. Mann of the Lick Branch mine, the site of two recent explosions. "You will observe," he said, "with what wonderful regularity they occur about the time the West Virginia legislature goes in session, and with what vigor and insistence the labor union lobby and all the labor union politicians insist upon liscensed [*sic*] miners, the prohibition of other labor than that of citizens of the State and all other kinds of fool things, basing all their claims [on the argument] that the ignorant laborer working in our mines is the cause of explosions and consequent loss of lives."[13]

Seven years brought no change in Collins's attitude toward the union. He could still write that "we should drive the Union out of the State from one end to the other, as far as the coal fields are concerned."[14] But a change had occurred in his attitude toward coal-mine explosions. He had absorbed the "new" scientific information indicating that dust explosions took place, and with greater frequency in cold weather than warm; cold weather, not the meeting of the state legislature, signaled the period of danger in the mines. In 1911 Collins wrote to his superintendent: "You should watch your No. 2 mine with great care and see that no dust is allowed to be made under any circumstances. No expense should be spared to keep the dust down and avoid the danger of an explosion. . . . If we should have an explosion at this time, on top of all our other troubles, we had about as well give up the fight."[15] Collins wanted the explosion problem solved not only because a solution would prevent "drastic" legislation, but also because his mining operation was in financial difficulty and could not withstand the burdens of a destructive explosion. Other correspondence indicates that Collins wanted his employees to have an adequate insurance system, that he took safety into consideration when contemplating the wisdom of speeding up a particular mining process, and that he was receptive to the criticisms of the state mine inspector. George Wolfe, general manager of Collins's Winding Gulf Company, could, evidently with impunity, write to Collins that he had stopped work in a particular section of the mine following an inspection, "as we do not care to take any chances in matters of this kind even through [*sic*] it curtails our tonnage for a few weeks," and Wolfe had sufficient confidence in Collins that he could only months later commit the company to a $3,000 expense for ventilation without consulting the owner. Collins did not in any way resist Wolfe's plan to put the mines of the Winding Gulf Colliery Company "in first class shape." Though he did not participate in the national campaign for the Bureau of Mines, Collins's commitment to safety—especially considering his hostility to unionization—appears to have been every bit as great as that of the more vocal Fleming. The greater subtlety of Fleming's politics should not obscure the essential similarities: both were businessmen, concerned with profit and production, accepting safety if it meant direct or indirect gain, rejecting safety in its more costly forms—especially state legislation.[16]

If one assumes that employer opposition to unionization was primarily economic in impulse, the economic model also accounts for the use of safety as an anti-union device. Hoping to allay worker discontent and undercut the union movement through various forms of welfare, employers provided housing, medical facilities, schools, recreation, and safer working conditions. Thus company safety work was often part of a larger welfare movement motivated not by selfless humanitarianism but by the desire to beat the unions at their own game.[17] The most common form of anti-union paternalism in the coal-mining industry was the employee representation plan, placed into operation where recognition of the union had been refused. The first such plan was established in the Colorado Fuel and Iron Company mines of John D. Rockefeller following the Colorado coal strike of 1913, and the model soon was adopted by other large companies, among them the Davis Coal and Coke Company of West Virginia and the Pacific Coast Coal Company of Washington. Under the Colorado plan, the company's twenty mines were divided into four districts. The employees of each district selected three of their number to serve on a joint committee on safety and accidents with three employees selected by the management. The joint committee could bring to a company official (starting with the superintendent) any matter pertaining to safety in the mines. Employees with grievances over safety could relay them through the joint committee which, with the company mine inspector, inspected the mines three times a year.[18] Historians of employee representation plans have found the reality less encouraging than the theory. Employees were reluctant to register their grievances with company foremen or superintendents for fear of being discharged; miners were highly critical of their representatives and claimed they did not fairly represent them.[19] That safety was a secondary consideration in such plans was made explicit by one Pennsylvania company official. "When government regulation of Coal Mining had to come in Western Penna.," he said, "we did our best to place it on a parental basis instead of a police basis desired by trade union demogogues [*sic*]."[20]

United States Steel Corporation, in the forefront of the movement to extirpate unions, also employed the most advanced safety measures and systems in its coal-mining subsidiaries. One contemporary who was close to all phases of the safety movement described the work of United States Steel's H. C. Frick Coal and Coke Company in terms which reveal the fruits of paternalism: "Many of this company's precautions against accidents are not prescribed by law, but are subject entirely of the company's own initiation and adoption. It has, in fact, anticipated every legal measure laid down by state or national government for mine safety."[21] The Frick Company also shares responsibility for the slogan "Safety First," having employed its predecessor, "Safety the First Consideration," in a 1907 safety campaign.[22]

United States Steel's interest in safety was also consistent with the company's size. The larger coal companies, with their superior financial resources and greater overhead, were usually the ones to institute employee representation and other programs of more exclusively safety content. Only the larger companies could consider placing telephones in the mines; finance their own scientific investigations of safety problems; send men to Europe to study modern methods of conservation and accident prevention, as the H. C. Frick Company did; or publish safety literature in several languages.[23] They were usually more aggressive in adopting permissible explosives and special sprinkling cars for the control of coal dust and in establishing company safety programs. "Go where you will among mining operations to-day," said H. M. Wilson to the first session of the National Safety Council, "and if the company is big enough you will find a safety organization."[24] When the Ohio Mining Commission suggested a costly system of safety foremen, one journal reported favorable reaction only from the large Pocahontas Company. The others, said *Coal Age*, are "not so economically favored."[25] In Arkansas, the small operators formed the core of opposition to shot-firing legislation, claiming it would put them out of business. Mine inspectors generally agreed that the larger companies used more specialized safety methods and more advanced safety materials than did the smaller companies and that they were less likely to evade the mining laws. "The greatest trouble we have," noted a Colorado inspector, "is with the small fellows."[26]

The association of safety and economics was reinforced during the Progressive years by the efficiency movement, as increasing numbers of businessmen came to equate safety with economy and with corporate and national efficiency. This tendency ran deepest within the National Safety Council, a businessman's organization, where H. J. Bell of the Chicago and Northwestern Railroad announced that "Safety is efficiency, and every time a man is injured so that he has to give up his duties for a time a new man has to be put in his place and the whole work is disorganized. . . . I think every foreman is interested in production, and if he is interested in production he is interested in efficiency, and, first of all, a foreman must be interested in Safety."[27] The efficiency theme took on added importance with the entrance of the United States into the European war; David Van Schaack, council president, called on his organization to promote "our national efficiency" by "saving thousands of men and women each year from accidental death and maiming."[28] The safety movement, said Ferd C.

Schwedtman of the National Association of Manufacturers, "is not only a humane movement, it is also a business proposition, a question of efficiency. . . . in making cripples of our workmen, . . . we are wasting from four to five times as much energy and money as it would cost to prevent them."[29] In coal mining, the operator commitment to safety as efficiency was less than complete, though there was a good deal of lip service to the concept and some genuine recognition of how costly mine disasters could be. The economic cost of mine fires was particularly well known to operators, and the Bureau of Mines was created partly to prevent this burning of corporate assets. The defensive nature of the operator approach to the efficiency question is indicated in this statement by a Pennsylvania operator: "I don't know why there always seems to be such an insinuation against that word 'economy,' as though mine operators considered economy before safety. Now, it was not considered in that light at all. Many of the big mine operators of to-day look on the words 'economy' and 'safety' as being synonymous terms."[30]

Between 1910 and 1920 operators came increasingly to view safety as a reform movement which made economic sense, and the basic reason was workmen's compensation legislation. Under the system of employers' liability, which was dominant well into the twentieth century, employers could avail themselves of three defenses under common law: fellow servant, assumption of risk, and contributory negligence. Although the fellow servant and assumption of risk doctrines were modified by state courts and legislatures in numerous states during the 1890s and early 1900s, few of those injured or killed—by one estimate 12 percent of the total—were compensated. The Ohio agency of the Aetna Insurance Company paid only 6 percent of its claims for injury and death from 1903 to 1910.[31] The historian of the Iowa system concluded that it did little or nothing to reduce the number of work accidents.[32] John Mitchell called the legal defenses available to the employer "decayed relics of the so-called wisdom of the law."[33] After 1900 a revolution took place in systems of compensation for industrial accidents. The first effective compensation act was a federal one, passed in May 1908 for the benefit of men injured or killed while working in hazardous occupations in service to the United States government. Montana's act of 1909 was the first to affect coal miners explicitly. The big boom came in 1911, when twenty-three states enacted some kind of compensation legislation or created commissions to study the problem. While none of these early laws made provision for compensation of occupational diseases, most of the statutes, using a phrase borrowed from the British compensation acts, provided compensation for "personal injury by accident arising out of and in the course of employment," and those of eight states, including Illinois and Ohio, granted additional compensation or additional rights of action "for injuries caused by the employer's violation of the safety acts or by his personal gross negligence or deliberate intention to cause the injury."[34] Among the organizations lending strong support to the campaign for workmen's compensation were the National Civic Federation (after 1909), the United Mine Workers (after 1910), and the American Mining Congress (after 1911). By 1920 every state but six, all in the South, had some kind of workmen's compensation.[35]

Although the motives for coal-operator support of compensation legislation are obscure, they no doubt were not much different from those of the rest of the business class: the prospect of eliminating conflict with labor engendered by litigation over industrial injury, the attraction of rationalizing and making predictable the costs of accidents, the gradual erosion of the legal protections under employers' liability, and the fear of labor radicalism.[36] If, however, accident prevention were considered only a residual benefit of compensation legislation, the system apparently resulted in safer working conditions. Statistics are too poor, and relevant factors too many, to allow any precise estimate of the impact of workmen's compensation on accident prevention, but contemporaries considered the system effective. "Compensation undeniably is followed by prevention," John Mitchell stated flatly. Workmen's compensation laws, added Carter Goodrich, "have done something toward 'making props cheaper than men.' "[37]

Workmen's compensation functioned through private companies or state agencies, some states, like Ohio, making it difficult for private firms to operate within their domains. The Associated Companies was a private organization that served companies located in some eight states, including Illinois, Indiana, Pennsylvania, and Kentucky, which had compensation legislation encouraging private insurance, and it offers an opportunity to assess the effectiveness of workmen's compensation on a geographically limited but fairly extensive scale. Formed in 1915 and 1916 under the auspices of the Travelers Insurance Company, Hartford, Connecticut, Associated consisted of ten stock insurance companies united in a cooperative agreement designed to spread the risks of coal mining and "to furnish the final answer to the claim that coal mining risks with their large collective hazards cannot be distributed by means of stock insurance." From the beginning, Associated sought close relationships with federal mine-safety experts, calling on the Bureau of Mines to provide an impartial chief inspector for the organization. Although there was some disagreement within the bureau about the requisite qualifications for such a position, Associated ultimately selected H. M. Wilson for his superior administrative ability.[38]

Associated's distinctive contribution to accident insurance was schedule rating, a particular kind of merit rating which involved the comparison of actual conditions in a particular mine with perfect conditions in a standard mine in a limited, relevant, geographical area. The system was designed to reward, through lower premiums, operators who conformed to the safety standards established by Associated. The primary bottleneck in the system was the determination of the requirements of a safe mine within a particular state, and in adopting a set of safety standards the Associated Companies made use of the work of organizations such as the National Fire Protection Association and the Underwriters' Laboratories. Naturally, however, the most important source of information for standards was the Bureau of Mines.[39] The cooperative relationship between the Associated Companies and the bureau was a close one, in spite of its informality. Bureau officials heartily approved of risk classification, confident it would contribute to adoption of advanced safety appliances. George Rice, chief of the bureau's Mining Division, was well aware of the influential role the bureau would play in the formation of standards and stood firmly behind the Associated's efforts. The inspections of the Associated Companies, he wrote, would provide the bureau with "a far more effective agency than . . . the casual visits of the state inspectors, because there would be a money influence at stake behind the insurance inspector's reports."[40] The first standards of the Associated Companies were actually prepared in consultation with the bureau and state mining officials. James E. Roderick, chief of the Department of Mines in Pennsylvania, was particularly hopeful. "If you can succeed in forcing compliance with these rules," he said, "you will accomplish what the Department of Mines has tried to accomplish for years, but without complete success." The addition of the Associated inspectors to the state's staff would, claimed Roderick, produce a substantial reduction in mine casualties.[41]

Under the system in its final form, each state was rated as to "cause" of fatality, the total adding up to 100 percent. "Safety Organization" and "Safety Measures" each comprised 20 percent of the total, with the remaining 60 percent allotted to technical causes, such as accidents from electricity, coal dust (5 percent), falls of roof, and haulage. Each of these groupings—no matter how small a percentage of the total—was subdivided into one hundred points. An operator who had done everything possible to prevent a coal-dust explosion, then, would receive one hundred points for that particular category, which would amount to 5 percent of his total score. Although the standards included broad, general statements (superintendents should encourage mine safety), each category was standardized to some extent, for example, by fixing the ratio of foremen and shot-firers to total employees and by defining the contents of a good safety organization. The system was an obvious attempt to objectify inspection.[42]

Operator response to the Associated program was immediate and positive in those states, like Kentucky and Pennsylvania, which required mining companies to maintain compensation insurance. In Kentucky the compensation law had not been in effect ten days before operators were pressuring the Bureau of Mines to provide additional rescue and first-aid training facilities. The company's standards, said Wilson, had encouraged Kentucky operators to earn the reduction in premium which would result from having their men trained in first aid.[43] The bureau had already spent considerable sums on rescue and first-aid work in Kentucky, expenditures, it reminded Wilson, which had aroused no particular enthusiasm among Kentucky operators. The prospect of the bureau providing additional facilities, moreover, came into direct conflict with the bureau's desire to see operators undertake their own training. George Rice saw Wilson's plan as impractical, noting "it would merely relieve the operators of expense" and speculating that Wilson had put too high a premium on rescue stations when fixing his insurance rates. Two years later bureau officials were still worried that the stimulus of the Associated Companies would produce a demand for more first-aid facilities than the bureau could provide.[44]

In Pennsylvania the Associated Companies was able to induce operators to establish three cooperative rescue and training stations with their own funds. Wilson projected the cost of the stations and estimated the premium rates, and Pennsylvania mine owners responded quickly. Wilson's company also encouraged the formation of mine-rescue and first-aid organizations, and Associated insurance inspections prompted companies to establish safety committees and to employ additional superintendents. Difficulties arose in Kansas, where intransigent operators only slowly improved their mines; in Ohio, where statutes made private insurance an impossibility; and in regions where operators felt Associated standards were particularly inappropriate.[45] In general, however, the Associated experience indicates that coal operators would respond to appropriate statutory and financial stimuli.

Associated worked its limited good in an industry that remained tragically disorganized, "too poor to fight, too cowardly or too virtuous to steal," as an operator attorney had eloquently stated.[46] "As we all realize," wrote the empathetic Holmes to his friend Fleming, "as anxious as the operators in this country are to do everything possible for safety and for clean mining, it is impossible for many of them to carry out such plans because of the exceedingly low prices of coal at the mines."[47] Fully convinced that excessive competition lay at the heart of the industry's problems, coal operators between 1890 and 1920 searched long and hard for mechanisms which would mitigate its impact. To equalize costs of production and distribution over competitive areas, they sought uniform state legislation, uniform conditions under the interstate agreement, and even some forms of national legislation; they tried fixing prices and selling through common sales agencies; and when these attempts at cooperation were rebuffed by the Justice Department, operators turned to corporate consolidation and ultimately to national politics. In November 1909 Illinois, Indiana, Ohio, and Pennsylvania operators called on President William Taft to secure his approval of changes in the antitrust laws which would permit the coal operators, as Holmes phrased it, "to 'get together' and arrange for a price on bituminous coal at the mine, such as would render possible mining with less loss of life and waste of coal." The president, however, reportedly "could see no escape from a continuance of the present system of vigorous competition." "This only confirms our fear," Holmes wrote Mitchell, "that he does not realize what this system is, nor what it is doing for the coal miner and for our coal resources."[48] Operators held out hope that Woodrow Wilson's Federal Trade Commission (FTC) would sanction basic structural readjustments in the industry, but here, too, they were to be disappointed, as the FTC refused to exercise prior approval over trade agreements and passively acquiesced only in weak statistical associations.

Lacking political support to counter its inherent centrifugal tendencies, the coal industry was consistently frustrated in its organizational efforts. Although the earliest coal associations date back at least to the 1870s, the first ones of importance developed in the 1890s. The Illinois Coal Operators' Association was formed in 1897 to deal with competitive conditions affecting the whole state and to negotiate with the union; the Indiana Operators Association was organized in 1900, the Southwestern Interstate Coal Operators Association in 1903. In 1916 there were more than twenty local, state, and regional associations of operators, and in 1925 there were more than forty. Whether located in union or nonunion coalfields, these associations usually had increased safety as one of their goals. One of the constitutional objectives of the Smokeless Coal Operators' Association, for example, was to provide "the best, safest and most approved means for mining of coal and the safeguarding of miners and others employees engaged in such pursuits."[49] Some, like the Smokeless Coal Operators' Association and the Illinois Coal Operators' Association, engaged in the politics of safety in state and national capitals, while others, like the regional Coal Mining Institute of America, were concerned with safety only in its technical aspects.[50] Local, state, and regional associations, however, could neither deal effectively with the basic problem of competition in a national market nor work productively toward the goal of uniform state legislation. Solutions in these areas required a national association of operators. Impetus for such an organization came from Illinois, the state which had earlier separated from the joint conference but which now found its competitive position eroding under the banner of individualism. Illinois, said Chicago's *Black Diamond*, "was pleading for some united action that would bring a harmonious result in the various states. She wanted to end the practice of the miners of dividing the operators into groups and whipping them piecemeal."[51] Not long after negotiations began in 1909, it became clear that not all operators saw Illinois's suggestion as benign. A. B. Fleming expressed the viewpoint of most West Virginia operators. "It seems to me," he said, "that it would be impossible for our West Virginia Association to become a member unless we intend to 'unionize' and recognize the United Mine Workers, as I suppose all will do who join the National Association." When Ohio and Pittsburgh region operators also proved unwilling, the movement collapsed, and the American Federation of Coal Operators remained a regional association.[52]

The failure to create a viable national association vitiated a possible source of national reform and a possible lever for elimination of destructive competition. Organizationally, it also served to shift operator interest in safety into national organizations with more comprehensive memberships—the American Mine Safety Association, the National Safety Council, and the American Mining Congress. As the coal-mine operators' national political arm, the American Mining Congress (see Chapter 2) was the most crucial organization and the most disappointing. After 1910 it reverted to its traditional emphasis on metal mining and played a major role in the Bureau of Mines's new western orientation. More rewarding was operator involvement in the American Mine Safety Association (AMSA), a first-aid and rescue organization established in 1912. The National Safety Council (NSC), which absorbed the AMSA and its functions in 1915, also originated in 1912, the brainchild of Lew R. Palmer, chairman of the Safety Committee of the Association of Iron and Steel Electrical Engineers. The original sixteen-member organizing committee included Holmes and Wilson of the Bureau of Mines. For most of the years before 1920, coal operators maintained a strong position within the Mining Section of the council, but after 1915 the relative unimportance of coal mining in the NSC's structure produced dissatisfaction. Operators also were concerned with the expense of council activities and hoped to see a mine-safety association which would derive its financial support not from corporations but from dues paid by individual operators and miners. Plans for such an organization were under way in late 1919.[53]

That coal operators, presiding over the nation's most deadly occupation, had to participate in the safety movement as members of organizations dominated by other industries and pursuing disparate interests, was symptomatic of the coal industry's organizational malaise. That malaise was in turn only a reflection of the industry's atomistic structure, a condition that permeated its every aspect, safety included. Surely coal mining would have been a killer under any industrial system, but industry economics insured that operators would view safety not as efficiency, but as an expense to be avoided. "In the matter of accidents," wrote John Mitchell, "it not infrequently happens that an ounce of prevention costs more than a pound of cure."[54]

CHAPTER

6

Coal-Mining Safety and the Progressive Period

RICHARD HOFSTADTER'S *Age of Reform* is perhaps best known for its attempt to link Progressive reform to a status revolution—the Progressives were "victims of an upheaval in status," seeking change less to remedy social conditions than to satisfy personal needs.[1] Beneath this interpretation lies a descriptive analysis of the Progressive years which is of even greater importance:

> Curiously, the Progressive revolt—even when we have made allowance for the brief panic of 1907 and the downward turn in business in 1913—took place almost entirely during a period of sustained and general prosperity. The middle class, most of which had been content to accept the conservative leadership of Hanna and McKinley during the period of crisis in the mid-nineties, rallied to the support of Progressive leaders in both parties during the period of well-being that followed. This fact is a challenge to the historian. Why did the middle classes undergo this remarkable awakening at all, and why during this period of general prosperity in which most of them seem to have shared? What was the place of economic discontents in the Progressive movement? To what extent did reform originate in other considerations?[2]

Aside from the implication that reform impulses logically arise from personal deprivation, Hofstadter makes the key point: there was nothing fundamentally wrong with the nation. The muckrakers "were working at a time of widespread prosperity, and their chief appeal was not to desperate social needs but to mass sentiments of responsibility, indignation, and guilt. Hardly anyone intended that these sentiments should result in action drastic enough to transform American society. In truth, that society was getting along reasonably well."[3]

Having argued that the society was fundamentally healthy, Hofstadter was led to seek an explanation for reform within the reformer rather than within the society. The result—a status interpretation of reform—has recently been subjected to scrutiny from the viewpoint of the modern social sciences. In an important article, historian David Thelen has argued that Hofstadter's status approach could not be confirmed through reference to sociology and psychology. Thelen calls for a return to history and chronology and emphasizes the 1893–1897 depression, which "vividly dramatized the failures of industrialism," as the critical reform-producing event. The depression, according to Thelen, created "a clear sense of priorities," generating support for tax reform and attacks on various forms of "corporate arrogance."[4] Applied to coal-mining safety, Thelen's chronology would seem unrevealing. The states remodeled their mine-safety legislation in the 1890s and again after 1905, but there was nothing resembling an industrial safety movement until 1907. Still, Thelen has turned us in the right direction, away from the subconscious and personal motivations of reformers and toward an appreciation of historical conditions as a critical ingredient in the reform impulse.

This was not, in short, a society that was "getting along reasonably well." In fact, as Henry Adams so vividly suggested,[5] it was a society dangerously out of control, or which conceived of itself in those terms. Thelen emphasizes the critical *consumer* issues—unsafe railway crossings, air and water pollution, political corruption—but Americans were also suffering in their *producer* roles. Hofstadter's sanguine commentaries stand in peculiar contrast to the society reflected in its industrial accident picture. Coal mining may have been the most deadly of American industries, but substandard conditions existed almost everywhere, as technology outstripped private and governmental controls. Train-wreck coverage was a standard feature in the magazines of the day. In the iron and steel industry, a 300-day worker had close to a one-in-four chance of death or disability. Monongah was only the quintessence of what B. O. Flower's *Arena* termed "the deep-seated dry-rot which permeates a large part of the corporation business of our time."[6] Accidents are hardly an incidental feature of industrial life; they indicate, perhaps more clearly than any other single measure, the extent to which a society is willing to sacrifice human life or is incapable of preventing human sacrifice though it may desire to do so. As of 1907, American society suffered from both maladies: it was at once at the apex of its brutality and possessed of an inflexible and unwieldy federal system which proved a major obstacle in the reform movement to come. The statistics for accidents and fatalities suggest that in 1907 American culture placed a greater value—relative to human life—on production and distribution than at any other time in its history. It was this emphasis, as yet little influenced by the "New Competition," which created Monongah.

The industrial-accident problem produced a broad-based reform movement, grounded in the industries that were among the first to feel the effects of economic growth—coal, railroading, iron and steel, agricultural machinery. International Harvester, United States Steel, the Chicago and Northwestern Railroad, and the H. C. Frick Coke Company all initiated safety programs in 1907 or soon thereafter. The steel industry was the scene of a particularly effective safety effort, made possible by the industry's structural characteristics (large units, restricted competition) and promoted by an aggressive group of engineers and by a management aware of the value of paternalism. Founded in 1907, the Association of Iron and Steel Electrical Engineers urged safety measures on the industry and was responsible for the establishment in 1913 of the National Council for Industrial Safety (later renamed the National Safety Council).[7] "Probably in no other industry," concluded the Bureau of Mines in an annual report, "have the results of the intensive campaign of accident reduction been more energetic and costly, than in steel-producing establishments."[8]

Interest in industrial safety yielded a congeries of safety organizations and institutions. The first national safety exhibition took place at the Museum of Natural History in New York in 1907, under the auspices of the American Institute of Social Science, and led to the incorporation, under 1911 New York State law, of the American Museum of Safety, which was intended primarily to educate employers and employees. The American Mine Safety Association and the Joseph A. Holmes Safety Association were concerned less with public interest in safety than with the relationship between safety and the production process. The former was essentially a first-aid and rescue group; the latter, formed from representatives of large national groups upon Holmes's death, was dedicated to stimulating mine safety through financial awards for contributions in the field. The Safety First Federation of America was founded in 1915 to coordinate the work of local public safety bodies in the fields of fire prevention and transportation. The industrial accident situation also co-opted a share of the energies of organizations with a variety of reform interests. The American Association for Labor Legislation (AALL) dealt with such issues as occupational disease, enforcement of labor laws, limitation of working hours for women, industrial hygiene, uniform accident statistics and reporting, and workmen's compensation. A broad-based reform group whose membership included John Mitchell, Jane Addams, and Samuel Gompers, the AALL focused on mining safety in 1911.[9] The National Civic Federation (NCF), founded in 1900 as a tripartite (business, labor, public) agency to facilitate cooperation and agreement between labor and business, confronted the safety question indirectly, through support of workmen's compensation legislation, and directly, by its advocacy of a federal mining bureau. The organization also maintained a Committee on Prevention of Mining Accidents, including Mitchell among its members. Nonetheless, safety was not a major concern of the NCF and its identification with business forced Mitchell's resignation in 1911.[10] Still another organization attentive to industrial accidents was the American Academy of Political and Social Science, which committed its entire annual meeting in 1911 to the hazards of modern industry.[11] The extent of government involvement in the safety movement is indicated in Safety First Week, held in February 1916 and sponsored by nine bureaus (including the Public Health Service, the Steamboat Inspection Service, the Bureau of Medicine and Survey of the Navy, and the Bureau of Mines) whose activities were primarily in the service of protection of life and property.[12]

The industrial-safety movement was related to Progressive concern for resource conservation. The problem resides in separating the genuine intellectual ties between safety and conservation from rhetorical or incidental ones. Expressions of some relationship between the two were common enough. Minnesota Governor Adolph O. Eberhardt expressed the general idea in 1910, when he argued that plant and animal diseases, sanitation and health inspection of homes and schools, and railroad, mine, and factory accidents all came properly within the scope of conservation.[13] Charles Van Hise held similar views. "The conservation of man," he said, "is one of the main purposes of government, of remedial legislation, of innumerable organizations, philanthropic and otherwise. The science of medicine, political economy, politics, and sociology are largely directed to that end."[14] The National Conservation Congress devoted its 1912 annual meeting to human resource conservation, and, with the approach of world conflict, numerous safety experts associated human resource conservation with the nation's manpower needs. "Safety First! Everywhere, that warning halts, that lesson guides, the inspiration stirs," said Edson S. Lott, president of the United States Casualty Company, in 1915:

> And now, in this second year of the world war, the President of the United States halts and guides and stirs us all with a paraphrase, finer in sentiment, nobler in inspiration. Said he, the other day: America first! America first! And, if perchance America must, too, some time fight, we shall want men—strong men, sound men, contented men. Shall we not help to conserve these future fighters now? What if, through our safety campaign, we can now reduce the workers killed or maimed each year by 10, nay 15 per cent! For the nation's welfare, let us make it five, yes, ten points more. This is our part in preparedness. For against the future need, Safety First and America first go hand in hand.[15]

The facile analogy between the two variants of conservation expressed a close relationship: human and resource conservation were separate aspects of the far-reaching Progressive concern with efficiency. For an important segment of the industrial-safety movement, human conservation was a means, not an end. Just as the Progressive conservation movement was less concerned with democratizing ownership of scarce resources than with efficient resource use, so too was the interest in human resources strongly (perhaps even primarily) influenced by concepts of efficiency.[16] According to Samuel Haber in *Efficiency and Uplift*, Progressives used the word efficiency in four ways: first, to depict a personal characteristic, related to a person's efficiency and effectiveness; second, to describe the energy input-output ratio of a machine; third, to denote the making of money (commercial efficiency); and fourth, to signify "a relationship between men. Efficiency meant social harmony and the leadership of the 'competent' " (social efficiency). Of the four, commercial efficiency and social efficiency were central to the movement for safety in the coal mines.

Commercial efficiency involved, on its most basic level, the application of specific processes to businesses in order to increase profits. Thus firms who found safety an efficiency concept, who saw in it profit possibilities, might institute safety programs, support safety legislation, or lend their support to the campaign for a bureau of mines. Safety engineer George Fonda of Bethlehem Steel perfectly illustrated the spirit of those who saw safety as efficiency. Having calculated the savings to be obtained from requiring employees to wear safety goggles, Fonda added: "Is it not true that 'Safety First' in many, many cases is synonymous with Efficiency Engineering?"[17] The attitudes of A. B. Fleming and Justus Collins, the impact of workmen's compensation legislation, and the operator response to the merit-rating program of the Associated Companies—all are indicative of the place of commercial efficiency in coal-mining safety.

But if commercial efficiency can claim much of the credit for business interest and participation in coal-mining safety reform, it must also shoulder some substantial portion of the blame for the movement's limited achievements. Commercial efficiency was a two-edged sword which more often cut against safety than for it. Because of the competitive structure of the industry and the marginal performance of many of its firms, commercial efficiency most often meant cost-cutting, trimmed expenditures for wages and safety. Its effectiveness as a device for the conservation of human resources was also seriously impaired by an unfortunate chronological coincidence: the industrial-safety movement began and reached its peak at a time when labor was plentiful and, in an economic sense, hardly in need of conservation. Unlike grazing land or timber, labor was not a disappearing resource.[18] Although unionization and licensing had begun to restrict the labor supply in the coal industry, there is no indication that operators were suffering from any general inability to secure workers. Until 1914, unrestricted European immigration insured an adequate supply of new immigrants to work the mines. The restricted flow of immigrants after 1914 resulted in a rhetorical connection between human conservation and manpower needs, but there is no evidence of a reinvigorated industrial-safety movement.

Mine workers were not only of no particular value economically; they were also of declining usefulness culturally, and this, too, must have limited the ability of the coal-mining safety movement to attract support and maintain enthusiasm. Again, a historical coincidence may lie at the center of the problem: the industrial-safety movement corresponded almost perfectly with a resurgence in American nativism and racism. At the same time that blacks and eastern and southern Europeans were replacing Scots, Welsh, English, and Irish as the dominant elements in the coal-mining work force, the new mining population was increasingly viewed with resentment and alarm: the new immigrants threatened American institutions of self-government, national strength, and unity, brought with them violence and disrespect for law, and, in the South, threatened to disrupt the racial status quo. The hostility with which these blacks and new immigrants were viewed by the American public—including the AF of L, politicians of the major parties, intellectuals and scientists—made unlikely a sustained social justice movement in their behalf and increased the intensity of ideas of social efficiency, harmony, and control within the industrial-safety movement.[19] The essence of social efficiency was the desire to eliminate sources of friction within the society and to reduce the frequency of undesirable labor actions—particularly strikes, unionization, and independent political activity. To achieve these goals, Progressive businessmen instituted private welfare programs in their plants and mines. These programs, incorporating safety, education, and recreation, formed what was known as the industrial betterment movement. The place of safety in the movement was explained by National Safety Council president David Van Schaack: "There can be no doubt of the great value of safety work, both in itself and in the stimulus which it inevitably gives to other branches of industrial betterment. . . . It points the way unfailingly to other fields of social well-being. It contributes to individual happiness, to the better understanding of man by man, to [the] spirit of fellowship."[20] Efforts to stabilize the business climate and to eliminate disruptive dissimilarities in society (e.g., those caused by the new immigration) also conditioned parts of the industrial betterment movement. If not totally grounded in cynical calculation, the movement was not an altruistic venture either. Its major impetus came not from the church but from the corporation; its major advocates were those who felt uncomfortable in the presence of the masses. As in parallel reform movements in prohibition and education, in coal-mining safety social service was a means to the end of social efficiency.[21]

The high value placed on order, stability, social integration, and cohesion at certain points in American history has been central to a number of recent studies. For the Jacksonian period, the works of David J. Rothman on asylums, Michael B. Katz on education, and Ronald G. Walters on abolitionism emphasize the critical place of social control in three areas of mid nineteenth-century reform.[22] Less work has been done on this theme for the years 1890 to 1920, but there is enough information to indicate that the Progressive years were a time of great anxiety, of obsessive fear of social disorder and dislocation. David Musto has recently pointed out the close association between Progressive narcotic reform and prejudice against Chinese,

Negroes, and aliens; Joel Spring has emphasized social unity as the central goal in educational reform; Michael Lesy's free-form *Wisconsin Death Trip* argues that conditions of life in American small towns and countryside created intense anxiety, high rates of insanity and suicide, and tendencies to obsessive-compulsive or paranoid personality types. Further evidence of a general cultural insecurity can be found in the growing literature on race and nativism.[23]

This perspective of fear and anxiety must have considerably enhanced the significance of mine explosions, train wrecks, and other signs of industrial decay. Americans might now be moved to consider if their system of production had not advanced beyond their ability to control it. The confluence of social anxiety and nativism had particular significance for the content of the coal-mining safety movement, since the industry was increasingly populated by Italians, Slavs, and Poles—the very ethnic groups which the society held responsible for some of its difficulties. As a result, the coal-mining safety movement was biased in its analysis of and prescriptions for the safety problem, casting the new immigrant miner in the central role—as the agent of disaster rather than its victim. The result was an unfortunate focus on the miner as the key human factor in accidents and fatalities, a viewpoint that led the safety movement to an excessive emphasis on remedying defects in the miner's educational background and cultural heritage. Even here the possibility of progress was cynically regarded by those observers who believed the problem to be essentially racial, requiring long-term evolutionary change. The widespread use of the word discipline to describe a solution to the problem of mine accidents is perhaps the best indication that mine accidents and fatalities were believed to be caused by an absence of such discipline, by the absence of order and control. Unfortunately this analysis, too, was thrown back on the miner, for discipline usually meant that the worker, through education, was expected to internalize mine-safety values. In general, because the mine-safety question was culturally defined, a number of promising areas, such as technology, supervision, and precise legal liability, were not adequately explored.[24]

Despite this undeniable bias, Progressives, whether operators, miners, inspectors, scientists, or humanitarians, maintained an essentially modern problem-solving orientation. Operator Glenn W. Traer expressed the Progressive conception of reform as an ongoing, indeterminate process when asked if he thought coal-mining safety required a permanent bureau. "Yes I do," he said. "The education on this subject will never be complete."[25] Coal-mining safety, like other Progressive reforms, was a process rather than an essence, a way of doing things as much as a final result. To staff their bureaus, Progressives relied on professionals—experts in administration, science, engineering, and statistics. This expertism was particularly relevant to coal-mining safety, where the problems to be solved required sophisticated analysis and where the temptations to view industrial accident prevention in moralistic terms were so great. The following comments by the *United Mine Workers Journal* (whether serious or ironic) indicate the pervasiveness of expertism in the Progressive period: "It is good-by, Tuberculosis. The statistician is at work on its corpse. If an abundance of complicated figures, compiled by the expert statistician will not kill off the white plague then all hope is gone."[26] Joseph A. Holmes, the administrator-scientist, almost a jack-of-all-trades in this era of specialization, presided over an organization of other administrators (Van Manning), scientists (George Rice), engineers (Clarence Hall), and statisticians (Albert Fay). The particular importance of engineering to the safety field was reflected in the new profession of safety engineering. F. E. Morris opened his address to the National Safety Council with "We of this profession—and it is a profession. . . ."[27]

Like a number of Progressive reform movements—in child labor, workmen's compensation, and pure food and drugs, for example—the movement for coal-mining safety on the national level contained a strong element of humanitarian Progressivism. Particularly in its early stages, the movement was buoyed by muckrakers, socialists, and publicists, groups whose interest in national reform—usually meaning a bureau or a department of mines—could not be measured in financial terms. This type of Progressive, as historian Russel Nye has written, wished to use the state positively, "to promote and protect the public social and economic welfare."[28] Socialist Victor L. Berger expressed the intensity of feeling of humanitarian Progressivism: "Men and women are killed in factories, on railroads and in mines because human life counts for less than do the products of labor. Under our civilization the dollar is of more importance than the man."[29] The *Pittsburgh Survey* also belonged to this brand of Progressivism, which was aided by the public nature of the mine fire and the mine explosion. Monongah and Cherry were more than disasters, they were national events, facilitators of emotional involvement in the cause of mine safety. The Cherry disaster, for example, pulled Graham Taylor and his *Survey* magazine into the safety movement.[30] Surprisingly, however, no liberal reform group emerged to shape this emotional potential and present it politically. Coal-mining safety had no equivalent of the National Child Labor Committee, the American Association for Labor Legislation, or the Social Reform Club, a circumstance that left labor to carry the banner of social justice in the reform process.[31]

The United Mine Workers, the mainstay of Progressive interest in safety in the states, also received the backing of urban immigrant lawmakers. In Illinois this meant support from Chicago's new immigrant groups for coal-mining safety legislation which would benefit miners downstate. "This is all the more significant," writes historian John D. Buenker, "since there was little self-interest involved here for Chicago lawmakers. Their support has to be attributed to a general sympathy for other disadvantaged groups and a belief in the principle of safety and welfare legislation."[32] Buenker's attempt to link coal-mining safety legislation with J. Joseph Huthmacher's urban lower-class interpretation of Progressivism deserves a skeptical appraisal, largely because it ignores important characteristics of the politics of mine safety. In Illinois, Pennsylvania, Ohio, and West Virginia, safety legislation was not solely or even primarily the product of traditional interest group politics, with urban immigrants joining labor to combat business influence. Instead, political differences were initially compromised by commissions of operators, miners, and public (usually chief inspectors) representatives. When the results of compromise were presented to the legislature, they were not controversial and, in fact, not essentially political. Placed in this context, the pro-safety votes of urban immigrant representatives become not so much an active element in the political process as a predictable reaction to a compromise which had already taken place.[33]

Implicit in the Huthmacher-Buenker scheme is a weakness common to most political history: a tendency to see government as a passive rather than an active element in reform. Recognition of the centrality of bureaucratic reform is particularly vital in areas such as safety, where legislation was complex and technical and expert counsel correspondingly compelling. In the states, chief inspectors usually initiated reform activity through criticism of existing law and recommendations for revision, participated as "experts" on the commissions which formulated legislation, and employed administrative discretion to fill gaps in the resulting legal edifice. Nationally, Joseph A. Holmes's role in the creation and modification of the Bureau of Mines is sufficient evidence of the critical importance of bureaucracy. The head of the Technologic Branch of the Geological Survey succeeded because, as a bureaucrat, he possessed essential knowledge of the little-known technical problems of mine safety and because, as a bureaucrat, he was able to gain influence in Congress. In his hands, first the Technologic Branch and then the Bureau of Mines not only participated in politics as interest groups but also reached out to miners, operators, inspectors, and conservationists, motivating some, educating others, monitoring and coordinating and (when possible) controlling the political process. Although Holmes had more than his share of energy and political savvy, the role of activist bureaucrat was less his personal creation than a phenomenon of a centralized, technical society. Dr. Harvey W. Wiley, chief of the Bureau of Chemistry, had preceded Holmes with parallel efforts in behalf of federal pure-food and drug legislation.[34] It is time for greater recognition that government bureaucrats create as well as administer, and that entrepreneurship may exist as well within a public agency as within a private firm.[35]

In contrast to the very limited interest in the historic relationship between bureaucracy and reform, in the past fifteen years scholars have produced a considerable body of literature dealing with businessmen and reform. Although historians writing in this vein recognize the existence of humanitarian Progressivism, they have, in general, argued that Progressivism was much more than an altruistic quest for social justice and much less than a great liberal triumph over conservative opposition. Progressivism was, they say, also (James Weinstein) and primarily (Gabriel Kolko) a successful attempt by the business community to use the governmental machinery, particularly the national machinery, to achieve its own ends. "In the current century," according to Weinstein, "particularly on the federal level, few reforms were enacted without the tacit approval, if not the guidance, of the large corporate interests."[36] With some logic, most of the work of these scholars has emphasized the business/corporate role in economic regulation. Kolko, for example, has concentrated on the efforts of corporations to achieve stability in their economic affairs, to eliminate the chaos resulting from excessive competition and undependable governmental intervention. In this view, federal regulation and Progressivism were not "a counterpoise to the power of private business," nor "the complaint of the unorganized against the consequences of organization," two expressions used by Richard Hofstadter.[37] The railroads, threatened by the localism of labor and the farm, initiated "a movement to establish stability and control within the railroad industry so that railroads could prosper without the fearful consequences of cutthroat competition." Business supported federal regulation in the railroad industry and in other fields as a bulwark "against state regulations that were either haphazard or, what is more important, far more responsible to more radical, genuinely progressive local communities" than national regulation. "National progressivism," states Kolko, "becomes the defense of business against the democratic ferment that was nascent in the states."[38] Other scholars have shown businessmen taking reform positions vis-à-vis the Iowa dairy industry, the federal antitrust laws, and World War I mobilization.[39] Robert Wiebe has produced an influential variant of this interpretation, arguing that although some businessmen supported most efforts at economic reform and regulation, "an examination of businessmen's reactions to the Progressive movement indicates that far from forming a cohesive group they differed widely over the proper solution to America's problems and expended a large portion of their energies in internal conflicts."[40]

Though the business approach to social and political reform has been explored less systematically and less explicitly, some contributions stand out. Weinstein and Hays have convincingly demonstrated that urban businessmen, concerned with the inefficiencies and democratic, decentralized characteristics of city government, pressed for elimination of ward representation and for new manager and commission systems of administration.[41] Of the Progressive social reforms, workmen's compensation has attracted the most attention and produced the most disagreement among historians. Weinstein and Roy Lubove have provided the major reinterpretation. Although both argue that workmen's compensation was a business reform, Weinstein, approaching the subject largely from the perspective of the National Civic Federation, has emphasized business desires for social cohesion, while Lubove sees "concrete, material advantages" as central to the business viewpoint. Robert Wesser presents the more traditional case of a divided business community, largely opposed to workmen's compensation, in his analysis of the campaign for workmen's compensation in New York State.[42] Here Wesser's emphasis is similar to that of Wiebe who, while acknowledging the influence of businessmen in economic reform, minimizes their interest in social reform. "Social insurance laws were an anathema," states Wiebe. "The only important contribution which businessmen made to the social welfare movement came as a by-product of their zeal for civic improvement."[43] The work of Oscar E. Anderson, Jr., and James H. Timberlake, historians, respectively, of the pure-food and drug and prohibition movements, lies in an intermediate position: businessmen were influential but not central to these reforms.[44]

The coal-mining safety movement was only one small facet of Progressive reform, hardly the basis for accepting one view of business/political relationships in the Progressive period and rejecting another. Yet any comprehensive evaluation of Progressive politics must at least take account of several major aspects of the politics of safety. Coal-mining safety was, above all, a business reform. For all the complexities and nuances of national politics, the Bureau of Mines, the major Progressive innovation in mine safety, would not have been established without the cooperation of coal and metal mine operators in the American Mining Congress. Holmes was a masterful organizer and a knowledgeable and judicious politician, but he lacked a substantial power base outside of the business community. Businessmen were also dominant in the frustrated but historically important uniformity campaign. Even in state politics, coal operators in major states demonstrated a certain flexibility—and the absence of their dominance—by agreeing to participate in the work of legislative commissions. In short, the emphasis of Gabriel Kolko and James Weinstein on the primacy of business in the reform process seems appropriate to the Progressive coal-mining safety movement.

This is not to say that Robert Wiebe's model of business disunity can be entirely dismissed; it must, however, be applied selectively. During the campaign the coal operators were able not only to unite among themselves but to cooperate with metal mine operators. Applied to the period from late 1907 through 1910, the Wiebe model seems particularly inappropriate. Once the bureau had been created, however, mining-industry unity proved evanescent; metal mine operators withdrew from the short-lived alliance of convenience and pursued their own interests in moving the Bureau of Mines into metal-mining areas. Divisions within the coal industry were also in part responsible for the very limited achievements in uniformity and standardization.[45]

The history of coal-mining safety reform also casts doubt on the assumption—which I suspect is widely held—that the relationship between business and government is essentially a function of the type of reform (i.e., economic or social) under consideration. This classification seems particularly dubious when applied to workmen's compensation and industrial safety. In each case reforms were designed to provide industry with the same kinds of services it might obtain from clearly economic measures like the Bureau of Corporations and the Federal Trade Commission (FTC)—stability, predictability, and security.[46] Obviously, the FTC held potential which the Bureau of Mines and uniform safety and workmen's compensation legislation did not; but that should not obscure the analogous purposes behind economic and social reform.

The analogy might be extended one step further, from ends to means, from the goals of stability, predictability, and security to the process, centralization. How does one explain the emphasis of the coal-mining safety movement on various national solutions? Why did industrial safety reform follow railroad, banking, and antitrust regulation in seeking a national focus of activity? In *The Search for Order*, Robert Wiebe has provided an overview of the problem. In the nineteenth century, states Wiebe, America was composed of "island communities"—small towns, parts of cities, villages; dispersed, separated, and relatively isolated. Around and within these island communities, the forces of nationalization, industrialization, mechanization, and urbanization were dramatically changing the society, yet to most people these themes "meant only dislocation and bewilderment. America in the late nineteenth century was a society without a core. It lacked those national centers of authority and information which might have given order to such swift changes." Most nineteenth-century organizations were collections of local associations, "designed partly to re-create and partly to protect a sense of community among its members." Nineteenth-century reform was itself localistic, oriented toward the states, and expressing, through antimonopoly, basic desires or community self-determination. "In no sense," states Wiebe, "did the reformers expect to realize their program by . . . constructing a huge apparatus of centralized direction."

The primary agent of change was a new middle class, numerically large in the cities and including professionals and specialists in business, labor, and agriculture, a class determined to transcend local ties and to replace or augment them with occupational connections beyond their immediate locales. "Consciousness of unique skills and function," states Wiebe, "characterized all members of the class. They demonstrated it by . . . an eagerness to join others like themselves in a craft union, professional organization, trade association, or agricultural cooperative." Much of Progressivism was the substance and result of the aspirations and achievements of this middle class, seeking an outlet for their talents and trying to "locate themselves within a national system." This class was both cause and effect, created by and creators of the new national industrial system and the national apparatus established after 1900 to service and regulate that system.[47] Samuel P. Hays described the same process in *Conservation and the Gospel of Efficiency* though, as the title of the book would indicate, Hays locates the critical determinants of change within technical rather than social frameworks. He states:

> These new forms of organization tended to shift the location of decision-making away from the grass roots, the smaller contexts of life, to the larger networks of human interaction. This upward shift can be seen in many specific types of development . . . [in] the upward shift in regulation of economic life from the state to the federal regulatory agency. These upward shifts did not arise out of new political theories or the inherent logic of a proper distribution of governmental powers, but rather from the fact that those who fashioned the new patterns of system and functional organizations sought a framework of decision-making consistent in scope and applicability with the scope of affairs they wished to control. . . . New contexts of human life had arisen, giving rise to new contexts of conditions to be controlled. Control now became a more elaborate process, involving measurement and prediction, reliance upon the experts who could develop and manipulate information, and techniques for shaping the course of events to reach predictable outcomes. . . . in each case the larger forces of economic life, with a scope far broader than cities, regions or states, sought a national, uniform context of action and a central point of decision-making which greatly limited the political variables to be controlled.[48]

The process Hays describes here was at work in almost every area of American life. Centralization took place, for example, in the Iowa dairy industry, with the consolidation of creameries and the beginnings of state inspection; in urban government, as businessmen sought an end to the ward system; and in meat packing, where federal regulation was part of the packers' search for foreign markets.[49]

In the nineteenth century, coal-mining safety was the concern of the counties and the states—island communities in Wiebe's framework —which responded to safety problems with legislation which, if not adequate, was structurally appropriate in the sense that it roughly paralleled the localistic market structure of the industry. By 1890, however, coal-mining firms were regularly servicing regional and national markets. Aided by declining transportation charges, low-cost producers, particularly in West Virginia, were selling in Pennsylvania, Illinois, Ohio, and Kentucky markets. Labor markets, too, were national and international, and the coal industry, now seriously affected by immigrants who had never mined coal and transients who often lacked familiarity with state regulations, found itself governed by inadequate state institutions. Well before 1900 bewildering explosions had raised questions about the technical conditions of mining—about mine gases, coal dust, machinery, humidity, and temperature, for example—which clearly transcended the abilities and resources of the states as well as their boundaries.

Between 1895 and 1910, these conditions, and others not related to safety, produced significant organizational changes in the coal industry. The United Mine Workers developed from insignificance into the largest union in the nation; the American Mining Congress was established and, after a decade as a metal-mining organization, absorbed the coal operators; the Mine Inspectors' Institute of America was born of the Monongah disaster. These organizations, and others of lesser importance for safety, were on the one hand examples of centralization and, on the other, the agents of centralization in coal-mining safety. Led by the coal operators, these groups had in common the recognition that national problems required national solutions. To solve the national problems posed by technology and interstate competition, they created the safety program of the Technologic Branch of the Geological Survey and finally the Bureau of Mines. When state safety legislation proved incompatible with national markets, the industry sought to nationalize the federal system through the mechanism of uniform state legislation.[50]

Although there is more than a little danger in emphasizing the strength of the centralizing and nationalizing impulses and the extent to which the society had been reconstructed by organizational change by 1900 or even 1920, historians of organization have not been inclined to examine or assess the equally critical forces of localism and decentralization. The concept of an organizational "revolution," traceable to the publication, in 1953, of Kenneth Boulding's *Organizational Revolution*, has been propagated by Hays and Wiebe, the two leading theorists of Progressive period organizational developments. Each, admittedly, has included within his work an assertion of the incomplete character of Progressive period organizational change. Reformers, according to Wiebe, "built no more than a loose framework, one malleable enough to serve many purposes . . . in the end they constructed just an approach to reform, mistaking it for a finished product." Hays suggests that "local and parochial social organization" remains potent even in the 1970s. Yet neither scholar is much interested in probing anti-central and anti-national impulses. Their caveats are more afterthoughts than essential elements in their history.[51]

This is not to deny the existence of organizational change from 1890 to 1920, nor even the accelerated pace of organizational development on the heels of the 1893 depression.[52] It is not the facts of the matter that are in question, but their meaning. Should the organizational successes of the labor movement after 1895 be considered part of an organizational "revolution," though some 90 percent of the American work force remained unorganized? If the turn-of-the-century merger movement, the corporate element in the organizational revolution, was of such crucial importance, how does one explain the continued business interest in political forms, like the Bureau of Corporations and the Federal Trade Commission, which would presumably satisfy organizational desires? Do the organizational manifestations of the coal-mining safety movement—the Bureau of Mines and uniformity, in particular—suggest profound change? If we conclude that developments in labor, business, and safety were less than revolutionary and less than complete, then the focus of historical inquiry also shifts. Rather than why was change so *extensive*, the essential question becomes, why was change so *limited?*

From this perspective, the coal-mining safety movement becomes a halting and incomplete expression of organizational change. After all, the movement produced no national regulatory mine-safety legislation, not even a provision for administrative mine-safety regulations emanating from the bureau. Lacking coercive authority, the Bureau of Mines could function only as an educational and scientific agency; regulatory and inspection functions remained in the states. Uniformity, whose promise of centralization seemed so great, proved to be an unrealistic goal and a timid device for bringing unity and discipline to an unruly federal system. A similar pattern exists in the area of child-labor reform. The national achievements of that movement were the Children's Bureau—as powerless as the Bureau of Mines—and two national child-labor laws, declared unconstitutional in the United States Supreme Court. National action was made difficult by the persistent opposition of Southerners—reformers and businessmen alike—who did not want federal regulation.[53] National workmen's-compensation legislation covered very few workers; almost all legislative successes came in the states. Compulsory health insurance, much less popular than workmen's compensation, was considered seriously only at the state level. Significant national regulation was achieved largely in those areas of social reform in which social control was the primary stimulus to action. And even in narcotics regulation and prohibition, two outstanding examples of social-control activity, the states were the primary locus of reform activity through most of the Progressive period.[54]

In *The American Partnership*, an influential book among American historians, Daniel J. Elazar suggests that a model of dual federalism —of rigid separation between state and national governments—is of little value in describing the government of the nineteenth and twentieth centuries. According to Elazar, the dynamic American society of the nineteenth century placed stress on its original federal structure, to the point where federalism ceased to be dual (with national and state governments operating in separate spheres) and became cooperative (with national and state governments working together in almost every area of activity). Elazar's framework is an important one, but of limited usefulness in analyzing the coal-mining safety movement, a field in which cooperative activity existed only at the periphery of the reform process. Part of the problem is that cooperative federalism as an intergovernmental mechanism was, as Elazar says, the product of westward expansion, of a distended, frontier society. Although Elazar properly projects cooperative federalism into the twentieth century, he fails to acknowledge that the problem of distension and the frontier, while not eliminated, had by 1890 been merged with the post-frontier conditions of concentration and association.[55] In the nineteenth century the central problem of politics had been to unite the American people and the states; when that problem approached solution late in the century, it was replaced by another: how to reconcile the economic unity produced by revolutions in transportation and communications with the centrifugal politics of the separate states. To solve that problem the Progressives turned from the developmental politics of cooperative federalism to the politics of standardization, uniformity, and nationalization.

Uniformity was an enormously popular approach to reform, one that attracted labor leaders and capitalists, Republicans and Democrats, radicals and conservatives, and one that played a role in numerous aspects of Progressive reform. The origins of uniformity apparently go back to 1857 and New York lawyer David Dudley Field, but there was little concerted action in pursuit of uniformity until the American Bar Association (ABA) took up the cause in the late 1870s. In 1889 the ABA helped organize the National Conference on Uniform State Laws, a group largely concerned with business and legal matters. At about the same time, political scientists John W. Burgess and Simon N. Patten were employing the opening volumes of the *Political Science Quarterly* and the *Annals of the American Academy of Political and Social Science* to indict the state system for its manifest failures. Uniformity received a new stimulus and entered a period of new growth in 1906, when Secretary of State Elihu Root, speaking before the Pennsylvania Society of New York, called for the states to recognize their interdependence: "Every State is bound to frame its legislation and its administration with reference not only to its own special affairs, but with reference to the effect upon all its sister States. . . . If any State is maintaining laws which afford opportunity and authority for practices condemned by the public sense of the whole country . . . that State is violating the conditions upon which alone can its power be preserved."[56] The next year, Charles A. L. Reed of the University of Cincinnati called for the establishment of a council of states to formulate standard measures and present them to the states for enactment. Reed's idea bore fruit in 1925, with the organization of the American Legislators' Association. In the meantime, however, there was activity on a number of other fronts. Three organizations—the National Conference of Commissioners on Uniform State Laws, the National Civic Federation, and the American Association for Labor Legislation—were prominent in uniformity campaigns, each with its own areas of interest. In addition, dozens of more specialized organizations pressed for uniform legislation in particular fields, including taxes, workmen's compensation, compulsory health insurance, divorce, drugs, securities, child labor, and conservation. Among the Progressive period organizations and groups interested in some form of uniform coal-mining safety legislation were the United Mine Workers, the Bureau of Mines, the American Mining Congress, the Mine Inspectors' Institute, the American Institute of Mining Engineers, and the American Association for Labor Legislation.[57]

Regardless of the specific area involved, uniformity reflected a growing recognition of the interrelated nature of American economic and technical structures; it was intended to serve as a mechanism for dealing with the fundamental problems of interstate competition. Interstate and interregional competition, particularly in cotton processing, drove child-labor reformers first in the direction of uniform legislation, then toward national legislation. "One of the strongest impressions resulting from a study of the [National Child Labor] committee," writes Jeremy Felt, "is that of employers and manufacturers pushed into employing children by costs, competitive pressure, a very human lack of imagination, and, of course, by ordinary greed. The exploiter of child labor for its own sake seldom existed; in his place stood a man not ordinarily given to philosophical worries about the future of the race but determined not to give his competitors an inch. The only fair way to deal with child labor was through federal legislation."[58]

Workmen's compensation became the subject of a relatively successful uniformity campaign when state legislatures balked at placing their own firms at a competitive disadvantage.[59] Elsewhere, competition produced a more subtle variation on uniformity, as states tailored their reform programs to fit interstate economics. Massachusetts lawmakers, anxious that further additions to the state's advanced social and economic legislation would corrode its competitive advantages, took a number of conservative positions, opposing a federal commission which would establish railroad rates, liberalizing the state's incorporation policies, resisting new legislation for the protection of women and children.[60] Business interests could also seek a competitive advantage through reform itself. State railroad regulation came to Alabama in part because the state's businessmen found it hard to compete with their counterparts in Georgia, where rates had been held down by a commission. A number of businessmen came to favor state prohibition legislation for the competitive advantage it would yield over businessmen operating in states without this aid to industrial efficiency; and to advocate national prohibition as an aid in meeting competition from abroad.[61]

If, as I argue here, uniformity deserves to be recognized as fundamental to Progressivism, then Progressivism itself deserves reevaluation. The prominence of a tactic so patently idealistic, so dependent upon interstate cooperation, indicates that Progressivism was profoundly conservative in methodology. The primary political impulse of the age was neither the New Nationalism of Theodore Roosevelt nor the New Freedom of Woodrow Wilson, but rather an amalgam of the two, characterized by the desire to nationalize reform while maintaining the primacy of the states within the governmental structure.

Progressivism only hinted at the nationalism of the New Deal, offering but suggestions of the future prominent role of the national government, and those primarily in economic regulation. The Progressive age was indeed constructing an organizational society, but it was doing so cautiously and in its own distinctive way.

List of Abbreviations

AALL	American Association for Labor Legislation
AIEE	American Institute of Electrical Engineers
AIME	American Institute of Mining Engineers
AMC	American Mining Congress
AMSA	American Mine Safety Association
CMIA	Coal Mining Institute of America
IWW	Industrial Workers of the World
MIIA	Mine Inspectors' Institute of America
NCF	National Civic Federation
NFPA	National Fire Protection Association
NSC	National Safety Council
TAIME	Transactions of the American Institute of Mining Engineers
UMWA	United Mine Workers of America
UMWJ	United Mine Workers Journal
USGS	United States Geological Survey

Notes

INTRODUCTION

1. In 1907 the Consolidation Coal Company and the Fairmont Coal Company were both owned in part by A. B. Fleming.

2. U.S., Bureau of the Census, *Historical Statistics of the United States, Colonial Times to 1957* (Washington, D.C., 1960), pp. 356, 360; George Korson, *Coal Dust on the Fiddle* (Hatboro, Pa., 1965, orig., 1943), p. 5; Morton S. Baratz, *The Union and the Coal Industry*, Yale Studies in Economics 4 (New Haven, Conn., 1955), pp. 39, 42, 43; Doris Drury, *The Accident Records in Coal Mines of the United States* (Bloomington, Ind., 1964), p. 56. For the effect of mechanization on coal-mining safety, see UMWA, *Proceedings of the Twenty-third Annual Convention*, held at Indianapolis, Jan. 16–Feb. 2, 1912 (Indianapolis, 1912), p. 610, and Daniel Harrington, "Effect of Mechanization of the Coal-Mining Industry upon the Frequency and Severity of Accidents," U.S., Bureau of Labor Statistics, *Bulletin* 536 (April 1931):184–85.

3. Under the screen coal system, miners were paid only for those pieces of coal which would not pass through a standard size screen. Payment rates under mine-run were usually less than under the screen system to compensate for the lower market value of fine coal.

4. William A. Sullivan, *The Industrial Worker in Pennsylvania, 1800–1840* (Harrisburg, 1955), p. 74; Andrew Roy, *A History of the Coal Miners of the United States* (Columbus, Ohio, n.d.), pp. 84, 85, 198, 202; Edward Pinkowski, *John Siney* (Philadelphia, 1963), pp. 51–54; and Alexander Trachtenberg, *The History of Legislation for the Protection of Coal Miners in Pennsylvania, 1824–1915* (New York, 1942), p. 24.

5. Trachtenberg, *Protection of Coal Miners;* Earl R. Beckner, *A History of Labor Legislation in Illinois* (Chicago, 1929); Owen F. Beal, *The Labor Legislation of Utah* (Logan, 1922); Katherine A. Harvey, *The Best-Dressed Miners* (Ithaca, N.Y., 1969); Drury, *Accident Records*, pp. 56–57; Roy, *Coal Miners*, pp. 202–3, 219–20, 302–4. The fact that the states legislated on the problem at an early date is ignored or forgotten by historian Carnes and labor organizer Mother Mary Jones (see Cecil Carnes, *John L. Lewis* [New York, 1936], p. 6; and [Mary Harris] Jones, *Autobiography of Mother Jones*, ed. Mary Field Parton [Chicago, 1925], p. 30). A "major" disaster is one in which five or more persons are killed.

6. Jack Rogers, "I Remember That Mining Town," *West Virginia Review* 15 (April 1938):205.

7. Roy, *Coal Miners*, pp. 401, 403; Frank John Bietto, "A Study of the Federal Government's Attempts to Promote Safety in the Bituminous-Coal Mines of

the United States" (M.A. thesis, Southern Illinois University, 1952), p. 29, table adapted from *Congressional Record*, 77th Cong., 1st sess., 1941, pp. 2236–37.

8. Drury, *Accident Records*, p. 92.

9. Holmes to Members of House Committee on Mines and Mining (Jan. 25, 1913), U.S., Bureau of Mines, Record Group 70, United States Bureau of Mines Records (hereafter referred to as RG 70), box 104, 710, 1913. The number "710" corresponds to a subject heading in the original Bureau of Mines filing system. Since these numbers also serve to pinpoint document location within boxes, they are used whenever possible.

10. Bureau of the Census, *Historical Statistics*, p. 356.

11. From Table 24, "Major disasters in bituminous-coal and anthracite mines of the United States, 1906 to December 31, 1947," U.S., Department of the Interior, Bureau of Mines, *Safety in the Mining Industry*, by D. Harrington, J. H. East, Jr., and R. G. Warncke, *Bulletin* 481 (Washington, D.C., 1950):32.

12. Ibid., p. 5.

13. Adapted from Table 16, "Underground fatalities at bituminous coal and lignite mines, by principal causes," ibid., p. 18.

14. See Table 1, Arno C. Fieldner, *Achievements in Mine Safety Research and Problems Yet to Be Solved*, U.S., Department of the Interior, Bureau of Mines *Information Circular* 7573 (June 1950):30. See also table, p. 35, in Crystal Eastman, *Work-Accidents and the Law*, vol. 2 of 6 vols., *Pittsburgh Survey*, ed. P. U. Kellogg (New York, 1910); and tables, pp. 142, 143, in S. O. Andros, "Coal Mining Practice in District IV," Illinois Coal Mining Investigations Cooperative Agreement, *Bulletin* 12 (Urbana, 1915). The causal factors in nonfatal accidents were similar to those for fatalities.

15. This material is adapted from William Graebner, "Great Expectations: The Search for Order in Bituminous Coal, 1890–1917," *Business History Review* 48 (Spring 1974):49–72.

16. D. W. Kuhn, "Sherman Anti-trust Law with Special Reference to the Coal Mining Industry," *Proceedings of the American Mining Congress* 14 (1911):264; hereafter referred to as AMC *Proceedings.*

17. Commonwealth of Pennsylvania, *Report of the Department of Mines of Pennsylvania*, Part 2: *Bituminous, 1910* (Harrisburg, 1911), p. 4.

CHAPTER 1

1. Gerald D. Nash, "Bureaucracy and Economic Reform: The Experience of California, 1899–1911," *Western Political Quarterly* 13 (Sept. 1960):678–91.

2. U.S., Congress, Senate, Committee of the Senate upon the Relations between Labor and Capital, *Report of the Committee of the Senate upon the Relations between Labor and Capital*, 48th Cong., 1885, 1:32.

3. U.S., Congress, House, Industrial Commission, *Report of the Industrial Commission on the Relations and Conditions of Capital and Labor Employed in the Mining Industry*, H. Doc. 181, 57th Cong., 1st sess., 19 vols., 1900–1902 (Washington, D.C., 1901), ser. 4342, 12:174. This was a joint commission of the House and Senate augmented by nine presidential appointees. It employed leading economists to investigate aspects of United States economic life.

4. Industrial Commission, *Report*, 12:190, 89 (Schluederberg), 56 (Mitchell), xxix.

5. See *Congressional Record*, 61st Cong., 2d sess., 1910, 45, pt. 6:6304. State and federal regulation for safety of steamboats is discussed in *Steamboats on the Western Rivers* by Louis C. Hunter (Cambridge, Mass., 1949), pp. 520–46, and in John G. Burke, "Bursting Boilers and the Federal Power," *Technology and Culture* 7 (Winter 1966):1–23.

6. Under the original act of 1891, the president could appoint a mine inspector, who reported to the governor of the Territory and the secretary of the interior. If safety provisions in territorial mines were found inadequate by the inspector, either the governor or the secretary could order changes, and if they were not made, he could close the mine. Each mine was required to have two outlets and to conform to specific ventilation requirements (*Congressional Record*, 57th Cong., 1st sess., 1902, 35, pt. 2:1128–29).

7. *Congressional Record*, 51st Cong., 2d sess., 1891, 22, pt. 4:3915, 3557, 3824; ibid., 57th Cong., 1st sess., 1902, 35, pt. 2:1127; pt. 7:7366; pt. 8:7432. A report discussing ventilation in the territorial mines was submitted on Jan. 24, 1902, by the House Committee on Mines and Mining under the title *Protection of Lives of Miners in the Territories* (U.S., Congress, House, H. Rept. 148, 57th Cong., 1st sess., 1902, ser. 4399, 1:1–2).

8. *Work-Accidents*, p. 34. The *Pittsburgh Survey* was a systematic, six-volume study of working conditions in the Pittsburgh area, produced under the auspices of the Russell Sage Foundation. Although this volume was published in 1910, research was done in 1907 and 1908; the report describes conditions in 1906 and 1907.

9. *United Mine Workers Journal*, May 31, 1900, p. 4 (hereafter referred to as UMWJ).

10. Jay Hambridge, "An Artist's Impressions of the Colliery Region," *Century Magazine* 55 (April 1898):822.

11. *Safety in the Mining Industry*, p. 18. Most basic disaster statistics used in this study are taken from Table 23, pp. 25–31 (1906–1947).

12. "A Thirty-Day Record," *Collier's* 38 (Feb. 16, 1907):16–17. Picture spreads were a typical way of dealing with accidents for the national magazines of this period. See also editorial, ibid., p. 11.

13. Carl Vrooman, "The Ultimate Issue Involved in Railroad Accidents," *Arena* 39 (Jan. 1908):19. Louis Filler, *Crusaders for American Liberalism* (Yellow Springs, Ohio, 1939, 2d ed., 1950), pp. 211–12. *World's Work*, described by historian Louis Filler as "honest, sincere, and always one step behind the muckrakers" (Filler, *Crusaders*, p. 165), carried a substantial, informative article on railroad accidents in its Nov. 1907 issue, placing the blame largely on American materialism (Edward Bunnell Phelps, "America's Lead in Railroad Accidents," *World's Work* 15 [Nov. 1907–April 1908]:9575–79). The strongly conservationist and pro-Republican *Outlook*, edited by Lyman Abbott, searched for the responsible element in a Feb. 15, 1908, editorial, but carefully avoided choosing sides between labor and management (*Outlook* 88 [Jan.–April 1908]: 335–36). *Leslie's*, a past supporter of water pollution legislation and food and drug legislation, first placed the blame on divisions between worker and employee, then on worker materialism (editorial for April 4, 1907, *Leslie's* 104 [Jan.–June 1907]:312). Later the magazine criticized slackened discipline among em-

ployees, said to be encouraged by reading muckraking literature (editorial, 105 [July–Dec. 1907]:362, and Filler, *Crusaders*, p. 212).

14. First three verses of "The Monongah Disaster," originally printed in UMWJ, Jan. 9, 1908, p. 4, reprinted in Korson, *Coal Dust*, pp. 267–68.

15. The cause of the explosion is discussed in the *Preliminary Report* of the committee investigating the disaster. (West Virginia, Legislature, *Report of Hearings before the Joint Select Committee of the Legislature of West Virginia, Appointed under Substitute for House Concurrent Resolution No. 5 and House Joint Resolution No. 19, to Investigate the Cause of Mine Explosions within the State and to Recommend Remedial Legislation Relating Thereto, Together with the Preliminary and Final Reports* [Charleston, 1909], p. 601; hereafter referred to as *W. Va. Hearings.*)

16. R. O. Nuzum to A. B. Fleming of Fairmont Coal Company (Dec. 11, 1907), A. B. Fleming papers, West Virginia University Library, Morgantown, "General Correspondence," box 36. Aretas Brooks Fleming was the eighth governor of West Virginia, elected in 1888. At the end of his term he resumed law practice in Fairmont. He served as president of the West Virginia Board of Trade beginning in Oct. 1907.

17. UMWJ, Dec. 12, 1907, p. 4, and editorial, Feb. 28, 1907, p. 4. The Naomi mine in Fayette City, Pa., exploded Dec. 1, 1907, killing 34 persons.

18. *W. Va. Hearings, 1909*, p. 601.

19. Filler, *Crusaders*, p. 212; *Leslie's* 105 (July–Dec. 1907):590 (editorial); 106 (Jan. 2, 1908):8 (picture story); *Collier's* 40 (Jan. 4, 1908):12–13; *World's Work* 15 (Nov. 1907–April 1908):9928–32.

20. For these and other references on the response of the press, see U.S., Congress, House, Committee on Mines and Mining, *Hearings to Consider the Question of the Establishment of a Bureau of Mines*, 60th Cong., 1st sess., 1908, pp. 123–25.

21. U.S., Department of the Interior, United States Geological Survey, *Coal-Mine Accidents: Their Causes and Prevention: A Preliminary Statistical Report*, by Clarence Hall and Walter O. Snelling, with an introduction by Joseph A. Holmes, *Bulletin* 333 (Washington, D.C., 1907), p. 3. In 1904 Congress appropriated $60,000 for coal testing at the Louisiana Purchase Exposition in St. Louis. One member of the committee appointed to set up the coal-testing plant was Joseph A. Holmes. After the exposition, the coal-testing plant continued to operate and formed the nucleus for what became officially the Technologic Branch in 1907 (A. Hunter Dupree, *Science in the Federal Government* [Cambridge, Mass., 1957], p. 280). See also Fred Wilbur Powell, *The Bureau of Mines* (New York, 1922), pp. 2–3; and U.S., Department of Interior, Bureau of Mines, *Fifth Annual Report of the Director of the Bureau to the Secretary of the Interior, for the Fiscal Year Ended June 30, 1915*, Van H. Manning, director (Washington, D.C., 1915), p. 2 (hereafter referred to as Bureau of Mines, *Annual Report*).

22. Diary entry, Dec. 3, 1908, James R. Garfield papers, Manuscript Division, Library of Congress, Washington, D.C., container 8; Coal Operators of West Virginia and Other States, *Proceedings of Meeting, January 8, 1908* (n.p., n.d.), p. 33.

23. Coal Operators' Meeting, 1908, *Proceedings*, p. 31.

24. Italics mine. Report from the Director of the U.S. Geological Survey to the Secretary of the Interior, Feb. 3, 1906, U.S., Department of the Interior,

United States Geological Survey, Record Group 57, Records of the U.S. Geological Survey, Geologic Division, Library of Congress, Washington, D.C. (hereafter referred to as RG 57), file 82, pp. 7–8. The records of the Technologic Branch were destroyed by fire.

25. *Fairmont Times*, Dec. 24, 1907, from the Monongah Rescue Records (a scrapbook), West Virginia University, Morgantown.

26. William N. Page, President, Soup Creek Colliery Company, Fayette City, W. Va., to A. B. Fleming (Dec. 14, 1907), Fleming papers, box 36.

27. UMWJ, Sept. 21, 1905, p. 6. Mitchell to Cannon (April 24, 1908), John Mitchell papers, Mullen Library, Catholic University, Washington, D.C., box 11, file 46.

28. Page was an acquaintance of A. B. Fleming. Coal Operators' Meeting, 1908, *Proceedings*, pp. 6, 42, and introduction.

29. Page to Fleming (Dec. 14, 1907), Fleming papers, box 36.

30. G. H. Caperton to Fleming (Dec. 17, 1907), ibid.

31. J. C. McKinley of the J. C. McKinley Coal and Coke Co., Wheeling, W. Va., to Fleming (Jan. 2, 1908), ibid.

32. Fleming wanted to publish a report defending national government work on mines, but Holmes restrained him for fear of upstaging supporter James A. Hemenway, senator from Indiana (Holmes to Fleming, March 11, 1908, Fleming papers, box 36; also Report dated March 3, in files for Jan. 1908, ibid.); Holmes to Fleming (March 11, 1908), ibid.

33. *Fairmont Coal Company, the Explosion at Monongah Mines Fairmont Coal Company*, by Frank Haas, supplementing the *Annual Report of Operation* for 1907, Fairmont Coal Company *Bulletin* 11, Dec. 20, 1908 (Fairmont, W. Va., 1908), p. 43.

34. Fleming to McKinley (Jan. 3, 1908), Fleming papers, box 36.

35. Holmes to Pinchot (Nov. 9, 1907), Gifford Pinchot papers, Manuscript Division, Library of Congress, Washington, D.C., box 100, filed under H.

36. *Congressional Record*, 60th Cong., 1st sess., 1908, 42, pt. 2:1447–48.

37. Ibid., pt. 7:6543, 6542, 6544; and 60th Cong., 2d sess., 1909, 43, pt. 4: 3274, 3275.

38. Dupree, *Science*, pp. 161, 158–59, 165.

39. U.S., Congress, House, Committee on Mines and Mining, *Bureau of Mines and Mining*, H. Rept. 1065, 47th Cong., 1st sess., 1882, ser. 2068, p. 1; *Congressional Record*, 55th Cong., 3d sess., Dec. 19, 1898, 32, pt. 1:287. U.S., Congress, Senate, Committee on Mines and Mining, *Division of Mines and Mining, United States Geological Survey*, S. Doc. 40, 55th Cong., 3d sess., 1898, pp. 1–5; *Congressional Record*, 56th Cong., 1st sess., 1900, 33, pt. 2:1664 (S.R. 3109); and (generally) RG 70, box 9, 1910–11.

40. 1908 House Bureau Hearings, Dec. 20, Dec. 20, and Dec. 21, 1908, respectively, pp. 123–25. The figure 1,000 comes from U.S., Congress, Senate, Committee on Mines and Mining, *Establishing Bureau of Mines in Interior Department*, S. Rept. 692 to accompany H.R. 20883, 60th Cong., 1st sess., 1908, ser. 5219, 2:17.

41. 1908 House Bureau Hearings, pp. 55, 65–66, 82. Other coal operators present at the hearings who had similar opinions were Benjamin M. Clark of the Rochester and Pittsburgh Coal Mining Company, Alexander Dempster, representing the Pittsburgh Bituminous Coal District (101), O. L. Garrison, President

of the Big Muddy Coal and Iron Company, St. Louis, Mo. (120), and F. P. Bayles, Superintendent of Carbon Coke and Coal Company, Cokedale, Colo. (120–21). The Illinois Coal Operators' Association, at their convention in Peoria, February 29, 1908, also endorsed a bureau of mines.

42. Holmes to Fleming (Feb. 4, 1909), Fleming papers, box 37; Fleming to Scott (undated 1909 letter), Fleming papers, Aug.–Dec. 1909 box; 1908 House Bureau Hearings, pp. 15–17.

43. 1908 House Bureau Hearings, pp. 4–5. On the McHenry bill, see *Congressional Record*, 60th Cong., 1st sess., 1908, 42, pt. 4:3526–28, 3529. See also Paul W. Pritchard, "William B. Wilson, Master Workman," *Pennsylvania History* 12 (April 1945):81–108.

44. 1908 House Bureau Hearings, p. 57.

45. Ibid., p. 113; Walsh to Chairman and Members of House Committee on Mines and Mining, *Congressional Record*, 60th Cong., 1st sess., 1908, 42, pt. 7: 6711; pt. 4:3917.

46. AMC *Proceedings* 7 (1904):12.

47. Edward W. Parker, "The Prevention of Mine Accidents," AMC *Proceedings* 9 (1906):234.

48. For the idea of mining as coordinate with agriculture, and therefore deserving of a department of mines, see AMC *Proceedings* 7 (1904):26–27; 8 (1905):10, 20; 9 (1906):44; and 10 (1907):22, 34.

49. 1908 House Bureau Hearings, p. 51; Callbreath to Fleming (March 24, 1908), Fleming papers, box 36; AMC *Proceedings* 13 (1910):77; and AMC *Proceedings* 16 (1913):58–59. The funds for maintenance of Callbreath's Washington office and for his salary were provided by the Sierra Madre Club of Los Angeles and from voluntary contributions from operators in the Southwest. Unfortunately, the records of the AMC, including those of Callbreath's Washington office, are not extant. The Washington office probably was opened in early Jan. 1910 (UMWJ, Jan. 20, 1910, p .18).

50. AMC *Proceedings* 7 (1904):74.

51. AMC *Proceedings* 8 (1905):55, 79; S. Rept. 692, 60th Cong., 1st sess., 1908, p. 18; AMC *Proceedings* 10 (1907):31, 110; and 14 (1911):44, 28. For other communications with Roosevelt, see AMC *Proceedings* 10 (1907):34, and 14 (1911):44, telegram dated Sept. 28, 1909, Spokane, Wash. The relationship between the AMC and officials in the Geological Survey was very close. In 1904, for example, the survey sent four representatives to the annual convention, and Joseph A. Holmes of the U.S. Coal Testing Plant was on the board of directors of the congress (AMC *Proceedings* 7 [1904]:51–55, 7).

52. AMC *Proceedings* 10 (1907):313.

53. *Congressional Record*, 60th Cong., 1st sess., 1908, 42, pt. 6:5433.

54. UMWJ, Feb. 13, 1902, p. 4. United Mine Workers of America, "Proceedings of the Meetings of the National Executive Board," June 25, 1907, meeting (in MS), pp. 18, 15. This source is also available in print. The manuscripts are located in the national headquarters of the UMWA, Washington, D.C. UMWA, *Proceedings of the Nineteenth Annual Convention* held at Indianapolis, Indiana, January 21–February 3, 1908 (Indianapolis, 1908), pp. 305–6 (hereafter referred to as UMWA *Proceedings*).

55. UMWA *Proceedings*, 1908, pp. 351–53; 1908 House Bureau Hearings, pp. 5–7, 26–29, 34–35; June 25, 1907, meeting of the UMWA National Executive

Board (in MS), pp. 70–72; Mitchell to Taft (Nov. 27, 1909), Mitchell papers, box 134, file 191.

56. William B. Wilson told the House Committee on Mines and Mining in 1908: "I am firmly now of the opinion that it should be in the Department of the Interior. I fear that if it is placed in the Department of Commerce and Labor it will simply be used for the purpose and the extension of trade and the promotion of trade rather than for the protection of life, limb, and property, as it really should be" (1908 House Bureau Hearings, pp. 50–51). Callbreath of the AMC also favored a bureau in the Department of the Interior.

57. UMWA *Proceedings*, 1910, p. 237; editorial, UMWJ, March 10, 1910, p. 4; and UMWJ, May 26, 1910, p. 2.

58. U.S., Congress, House, debate May 21, 1908, *Congressional Record*, 60th Cong., 1st sess., 1908, 42, pt. 7:6711, 6713.

59. 1908 House Bureau Hearings, p. 73.

60. S. Rept. 692, 60th Cong., 1st sess., 1908, p. 21. See also the unpublished report by mining engineer John W. Groves, dated Nov. 9, 1908, based on visits to 58 western coal mines, in RG 70, box 38, 616.4, 1910–11 (hereafter referred to as the Groves report). *Congressional Record*, 61st Cong., 2d sess., 1910, 45, pt. 1:993; 60th Cong., 1st sess., 1908, 42, pt. 7:6722; and Holmes to Pinchot (Nov. 9, 1907), Pinchot papers, box 100, filed under H.

61. See Chapter 3 for a full development of this important Progressive concept.

62. House, debate May 21, 1908, *Congressional Record*, 60th Cong., 1st sess., 1908, 42, pt. 7:6709.

63. 1908 House Bureau Hearings, p. 85; and UMWJ, Dec. 2, 1909, p. 1, statements of G. W. Traer, Illinois operator, and T. L. Lewis of the UMWA; S. Rept. 692, 60th Cong., 1st sess., 1908, p. 22. See Mine Inspectors' Institute of the United States of America, *Proceedings*, held at Indianapolis, June 1908 (n.p., n.d.), pp. 14, 27; and *Proceedings*, held at Scranton, Pennsylvania, June 1909 (Pittsburgh, n.d.), pp. 5, 8, 15 (hereafter referred to as MIIA *Proceedings*). Holmes again was in attendance at the 1909 meeting.

64. Pinchot papers, box 3314.

65. 1908 House Bureau Hearings, p. 9.

66. Memoranda, Feb. 3, 1906, USGS Director to Secretary of the Interior, RG 70, file 82; April 2, 1906, USGS Director to Hon. James A. Tawney, House of Representatives, RG 70, file 125; and *Congressional Record*, 60th Cong., 1st sess., 1908, 42, pt. 7:6721.

67. Request of Sept. 8, 1907, quoted by Smith in Smith to J. A. Holmes, Chief Technologist, USGS (Jan. 28, 1910), RG 70, file 82.

68. The conflict between Smith and Holmes was complex and touched on issues of conservation and the proper function of the Technologic Branch. A similar interdepartmental problem may have had some influence on the possibilities for a bureau in 1906 and in later years. The General Land Office worked against measures which attempted to take power from their office. "If there should be an attempt to create a Bureau of Mines at the present time, much less a Department of Mines, it would probably meet with antagonism in certain quarters, notably in the Land Department," wrote D. M. Barringer in a Nov. 13, 1906, letter to a New York member of the AMC, James Douglas (RG 70, box 9, 1910–11).

69. *Leslie's Illustrated Weekly* 114 (June 20, 1907):588. See also speech by Douglas, May 21, 1908, *Congressional Record*, 60th Cong., 1st sess., 1908, 42, pt. 7:6707.

70. *Congressional Record*, 60th Cong., 1st sess., 1908, 42, pt. 7:6708.

71. Heyburn to Dick, March 21, 1910, Charles Dick papers, Ohio State Historical Society, Columbus; *Congressional Record*, 60th Cong., 2d sess., 1909, 42, pt. 4:3755, 3756.

72. Meeting of Coal Operators, 1908, *Proceedings*, pp. 9–10.

73. Clarence Hall, "Governmental Investigation of Mine Accidents," MIIA *Proceedings*, 1908, p. 36; *Congressional Record*, 60th Cong., 1st sess., 1908, 42, pt. 7:6714, 6709; Coal Operators' Meeting, 1908, *Proceedings*, p. 12. For the text of H.R. 20883, a composite of earlier bills on the subject, see *Congressional Record*, 60th Cong., 1st sess., 1908, 42, pt. 7:6705–6.

74. S. Rept. 692, 60th Cong., 1st sess., 1908, p. 26.

75. *Congressional Record*, 60th Cong., 2d sess., 1909, 43, pt. 4:3757; Mitchell to Andrew Roy (June 8, 1908), Mitchell papers, box 38, file 270, and William B. Wilson to Fleming (June 2, 1908), Fleming papers, April–Aug. 1908 box.

76. *Congressional Record*, 60th Cong., 2d sess., 1909, 43, pt. 4:3758, 3756–57; 60th Cong., 1st sess., 1908, 42, pt. 7:6720.

77. Holmes to Fleming (Feb. 9, 1909), Fleming papers, box 37. For further evidence of Henry Teller's responsibility here, see Englebright, "A Federal Bureau of Mines," AMC *Proceedings* 12 (1909):166; and J. W. Paul to Holmes (Feb. 22, 1909), RG 70, box 35, 1910–11.

78. AMC *Proceedings* 12 (1909):30.

79. UMWJ, March 18, 1909, p. 4.

80. Eastman, *Work-Accidents*, chapt. 3.

81. Germer to Debs (Dec. 8, 1909), and Debs to Germer (Dec. 15, 1909), in Adolph Germer papers, State Historical Society of Wisconsin, Madison, Box 1, file "Correspondence, 1901–1910."

82. *Congressional Record*, 61st Cong., 2d sess., 1910, 45, pt. 6:5957.

83. U.S., Congress, Senate, *Bureau of Mines*, S. Rept. 353, to accompany H.R. 13915, from Committee on Mines and Mining, Senate, 61st Cong., 2d sess., 1910, ser. 5583, pp. 1–2.

84. UMWJ, May 19, 1910, p. 3. See U.S., Congress, House, Bureau of Mines in the Department of the Interior, conference report submitted by Geo. F. Huff, H. Rept. 1291, 61st Cong., 2d sess., 1910, ser. 5593, pp. 1–2; *Congressional Record*, 61st Cong., 2d sess., 1910, 45, pt. 6:5954–55, 6035, 6037, 6040, and pt. 1:1001–2; debate May 3, 1910, *Congressional Record*, 61st Cong., 2d sess., 1910, 45, pt. 5:5644, 5645; and UMWJ, April 14, 1910, p. 3.

85. This sketch of Holmes is compiled largely from two sources: a sketch written in Aug. 1953, U.S., Department of the Interior, Bureau of Mines, Office of Minerals Reports (typewritten), and currently in their and the author's possession; and a sketch of Holmes by R. H. Sykes in the Joseph Austin Holmes papers, Southern Historical Collection, University of North Carolina, Chapel Hill.

86. Samuel P. Hays, *Conservation and the Gospel of Efficiency*, Atheneum ed. (New York, 1969), pp. 84–85, 131. Hays says: "As chairman of each of the Commission's four sections—Waters, Forests, Lands, and Minerals—the President chose either a United States Senator or Representative. But, for the sections'

secretaries, the members who carried the brunt of the work, he selected four tried and true conservation stalwarts, WJ McGee, Overton W. Price of the Forest Service, George Woodruff of the Department of the Interior, and Joseph A. Holmes of the Geological Survey."

87. J. H. Wheelwright, Vice President of Consolidation Coal Co., to Fleming (May 19, 1910), Fleming papers, box 38.

88. George W. E. Dorsey to Taft (Oct. 12, 1910), William Howard Taft papers, Presidential Series 2 or 6 (the first number refers to number in the index, the second to the box number; the index number will be used here), file 304, Manuscript Division, Library of Congress, Washington, D.C.

89. Taft to Ballinger (Sept. 3, 1910), Ballinger papers, microfilm reel 8. On microfilm, Manuscript Division, Library of Congress, Washington, D.C.; the originals are in the University of Washington Library, Seattle.

90. Gifford Pinchot, *Breaking New Ground* (New York, 1947), p. 56.

91. Editorials, UMWJ, July 12, 1906, p. 4; Dec. 17, 1908, p. 4; June 18, 1908, p. 4.

92. Roy to Mitchell (May 27, 1908), Mitchell papers, box 38, file 270.

93. Editorial, UMWJ, Feb. 3, 1910, p. 12.

94. *Mines and Minerals* 31 (Oct. 1910):162–63.

95. Editorial, UMWJ, Feb. 3, 1910, p. 12.

96. MIIA *Proceedings*, 1910, p. 34; Frank Haas, Consulting Engineer for Consolidation Coal Company, Fairmont, to Fleming (July 2, 1910), Fleming papers, July–Dec. 1910 box.

97. The latter possibility is mentioned in UMWJ, Sept. 28, 1911, p. 6, letter from Joe King to UMWJ, noting "that there was no harmony between the state and federal force of experts."

98. Holmes to Mitchell (Aug. 19, 1910, personal), Mitchell papers, box 15, file 63.

99. Laing to Fleming (July 5, 1910), Fleming papers, July–Dec. 1910 box.

100. Ibid.; UMWJ, July 29, 1909, p. 4, March 30, 1911, p. 7; Fleming to Laing (July 2, 1910), Frank Haas to Fleming (July 2, 1910), and Laing to Fleming (July 5, 1910), Fleming papers, July–Dec. 1910 box; Smith to Holmes (Jan. 28, 1910), RG 57, file 82.

101. Memorandum given personally by Garfield to Walter G. Fisher (March 28, 1911), James R. Garfield papers, box 128, last file, entitled "Official Correspondence, 1911." The same memorandum gives Garfield's opinion of Smith: "I selected Smith after full conference with Mr. Walcott and others in the Survey. I am now convinced that the appointment was a mistake; he has not proved as big and broad a man as I had hoped, and I feel from what I have heard that the Survey is not as strong as it should be." Garfield's original choice for the survey directorship was conservationist Charles Van Hise, president of the University of Wisconsin.

102. *Coal and Coke Operator* 11 (Aug. 4, 1910):494.

103. Reprinted in ibid. (July 28, 1910):479 and (Nov. 24, 1910):751.

104. Smith to Ballinger (July 5, 1910, official), RG 70, box 11, 1910–11. See also Guy Elliot Mitchell, "The New Bureau of Mines," *World To-Day* 29 (July–Dec. 1910):1150–55, esp. p. 1152, and Smith to Ballinger memorandum (Aug. 4, 1910), RG 70, box 11, 1910–11.

105. See Fleming to Laing (July 2, 1910) and reply (July 5, 1910); Smith to

Ballinger (July 6, 1910), Taft papers, Presidential series 2, file 669; Smith to Mitchell (June 14, 1910), Mitchell papers, box 11, file 46. Mitchell refused to support Parker because he had already expressed confidence in Ross and Holmes (Mitchell to Smith, June 30, 1910). *Coal and Coke Operator* 11 (Aug. 4, 1910): 494.

106. Smith to Holmes (Sept. 20, 1910), copy to Taft, Taft papers, Presidential series, 2, file 26; Smith to Ballinger (July 6, 1910), Taft papers, Presidential series 2, file 669; Donald Carr, Ballinger's secretary in Washington, D.C., to Ballinger (Sept. 2, 1910), Ballinger papers, microfilm reel 3; Carr to Ballinger (July 19, 1910), Taft papers, Presidential series 2, file 669.

107. UMWJ, Sept. 3, 1910, p. 1; Carr to Ballinger (Sept. 3 and July 26, 1910), Ballinger papers, microfilm reel 3. There is no indication that Holmes's conservation views had anything to do with the appointment controversy.

108. Carr to Ballinger (Sept. 3 and July 26, 1910), Ballinger papers, microfilm reel 3.

109. "Other bureaus" include, for example, the Geological Survey. Holmes to Ballinger (Sept. 6, 1910), Ballinger papers, microfilm reel 5.

110. Holmes to Mitchell (Aug. 19, 1910, personal), Mitchell papers, box 15, file 63.

111. Carr to Ballinger (July 19, 1910), Ballinger papers, microfilm reel 3; handwritten addition to a typed letter, Taft to Ballinger (Sept. 3, 1910), Ballinger papers, microfilm reel 8.

112. Ballinger to Taft (Sept. 13, 1910), Ballinger papers, microfilm reel 8.

113. Carr to Ballinger (Sept. 2, 1910), Ballinger papers, microfilm reel 3. Carr said: "We have not opposed Holmes to the extent of it being material to us whether he were appointed or not."

114. Scott to Taft, personal note (Sept. 4, 1910), Taft papers, Presidential series 2, file 1198.

CHAPTER 2

1. Bureau of Mines, *Sixth Annual Report* (1916), p. 41; *Second Annual Report* (1912), pp. 29–30; *Third Annual Report* (1913), p. 51.

2. Unpublished report entitled "A Syllabus of Problems in the Coal Industry Which Need Investigation," compiled by E. A. Holbrook, dated Jan. 10, 1920, RG 70, box 426, 101.5, 1919.

3. RG 70, box 19, 1910–11.

4. Ludwig Teleky, *History of Factory and Mine Hygiene* (New York, 1948), p. 247.

5. J. H. Bramwell, Stuart M. Buck, and Edward H. Williams, Jr., "The Pocahontas Mine-Explosion," *Transactions of the American Institute of Mining Engineers* 13 (Feb. 1884–June 1885):248–49 (hereafter referred to as TAIME); Roy, *Coal Miners*, pp. 219–20.

6. Roy, *Coal Miners*, p. 224.

7. E. S. Hutchinson, "Notes on Coal-Dust in Colliery Explosions," TAIME 13 (Feb. 1884–June 1885):278.

8. "Coal-Dust in Mine-Explosions," TAIME 24 (1894):898–917; George S.

Rice, "Investigations of Coal-Dust Explosions," TAIME 50 (Oct.–Feb. 1914): 554–55; and UMWJ, March 8, 1900, p. 1.

9. Holmes to Fleming (March 19, 1909), Fleming papers, box 37; Fleming to J. C. McKinley (Jan. 3, 1908), ibid., box 36; and Collins to A. M. Herndon, Superintendent of Winding Gulf Colliery Co. (Nov. 20, 1911), Winding Gulf Colliery Company papers, West Virginia University Library, Morgantown, box 1; Interstate Joint Convention, *Proceedings*, 1908.

10. Page to Fleming (Dec. 14, 1907), Fleming papers, box 36. The chief mine inspector of West Virginia at this time was James W. Paul who was, as Page said, a fanatic on coal dust. Paul also was proved right.

11. Italics mine. Page, "The Explosion at the Red-Ash Colliery, Fayette County, West Virginia," TAIME 30 (Feb.–Sept. 1900):862.

12. Coal Mining Institute of America, *Proceedings*, 1903–1904–1905 (Pittsburgh, 1906), Dec. 1904 meeting, p. 176.

13. Editorial, UMWJ, Jan. 30, 1902, p. 4. See also May 29, 1902, p. 4, and July 31, 1902, p. 5.

14. Editorial, UMWJ, March 2, 1905, p. 4.

15. Roy, *Coal Miners*, p. 229.

16. J. Taffanel, "Coal-Dust Explosion Investigations," TAIME 50 (Oct.–Feb. 1914):589. See also Teleky, *Factory and Mine Hygiene*, p. 248, and Rice, "Investigations of Coal-Dust Explosions," p. 555.

17. Editorials, UMWJ, Feb. 20, 1908, p. 4, and Feb. 28, 1907, p. 4.

18. Interstate Joint Convention, *Proceedings*, 1908, p. 67; UMWJ, Feb. 21, 1907, p. 4; Feb. 20, 1908, p. 1.

19. Holmes to Dick (Aug. 17, 1908), RG 70, box 1, 1910–11; Holmes to J. C. Kilson, President, Indiana Coal Operators' Association (Aug. 4, 1908), RG 70, box 1, 1910–11; and U.S., Department of the Interior, USGS, *The Prevention of Mine Explosions: Report and Recommendations*, by Victor Watteyne, Carl Meissner, and Arthur Desborough, *Bulletin* 369 (Washington, D.C., 1908).

20. Holmes to Meissner (Jan. 23, 1909), RG 70, box 1, 1910–11.

21. *Bulletin* 1 of the Juanita Coal and Coke Co., dated Sept. 1, 1907, in RG 70, box 35, 474, 1910–11; UMWJ, July 18, 1907, p. 1; for Illinois, see Beckner, *Labor Legislation*, p. 363.

22. UMWJ, Dec. 24, 1908, p. 7.

23. Rice, "Investigations of Coal-Dust Explosions," pp. 573–74; Bureau of Mines, *Third Annual Report* (1913), p. 46; *Fourth Annual Report* (1914), p. 43; *Fifth Annual Report* (1915), p. 37.

24. *Coal Age* 12 (Aug. 11, 1917):241; Rice, "Investigations of Coal-Dust Explosions," pp. 573–74.

25. Albert H. Fay, "Mine Accidents and Uniform Records," *Proceedings of the Second Pan American Scientific Congress* 8 (1915–1916):495; U.S., Congress, House, *Estimate for Investigation of Mine Accidents, etc., by Geological Survey*, H. Doc. 523, 60th Cong., 1st sess., 1908, ser. 5375, pp. 4–5.

26. Unsigned [probably from Holmes] to Meissner (March 17, 1908), RG 70, box 1, 1910–11.

27. Powell, *Bureau of Mines*, p. 10; Bureau of Mines, *First Annual Report* (1911), p. 54.

28. UMWA *Proceedings*, 1910, pp. 555–57; 1914, p. 1128; UMWJ, Oct. 21,

1909, p. 1; UMWA, Minutes of the National Executive Board, March 31 and June 25, 1909. This subject is explored in more detail in Chapter 4.

29. U.S., Congress, House, Committee on Mines and Mining, Mining Experiment Station in Colorado, *Hearings on H.R. 11414*, 62d Cong., 2d sess., 1911, p. 28; Bureau of Mines, *Third Annual Report* (1913), p. 16; Fay, "Mining Accidents and Uniform Records," pp. 495–96; John B. Andrews, "Needless Hazards in the Coal Industry," *Annals of the American Academy of Political and Social Science* 111 (Jan. 1924):27; Edward T. Devine, *Coal* (Bloomington, Ill., 1925), p. 235.

30. U.S., Department of the Interior, Bureau of Mines, *First National Mine-Safety Demonstration, Pittsburgh, Pennsylvania, October 30 and 31, 1911*, by Herbert M. Wilson and Albert H. Fay with a chapter on the explosion at the experimental mine by George S. Rice, *Bulletin* 44 (Washington, D.C., 1912), pp. 27–28; Bureau of Mines, *Fourth Annual Report* (1914), p. 17; Powell, *Bureau of Mines*, pp. 13–14; and RG 70, box 27, 1910–11.

31. U.S., Congress, House, Committee on Mines and Mining, *Appropriations for Mining Schools*, Hearings before the Committee on Mines and Mining, House of Representatives, on H.R. 6063, 63d Cong., 2d sess., 1913, p. 9; and Bureau of Mines, *First Annual Report* (1911), p. 33.

32. H. M. Wilson, Chief Engineer, to George S. Rice, USGS (Nov. 5, 1909), RG 70, box 134, 444.3, 1914; E. L. Carpenter to Holmes (April 1, 1915), Thomas J. Walsh papers, box 193, Manuscript Division, Library of Congress, Washington, D.C. See also E. D. Logsdon, President, Indian Creek Coal and Mining Co., Bicknell, Ind., to H. M. Wilson (Jan. 15, 1914), RG 70, box 105, 002, 1914; E. Drennen, Vice President and General Manager of Stonega Coal and Coke Co., Big Stone Gap, Va., to Holmes (June 4, 1915), Walsh papers, box 193; Ryan to Manning (Aug. 4, 1913), RG 70, box 88, 1913; RG 70, boxes 94, 445.42, 1913, and 360, 131, 1918; Ryan to Manning (Sept. 25, 1916), RG 70, box 238, 711, 1916; memorandum, James R. Garfield to Walter G. Fisher (March 28, 1911), Garfield papers, box 128, last file, "Official Correspondence 1911."

33. Memorandum, C. A. Barnard of the Owl Creek Coal Co. (Jan. 14, 1917), RG 70, box 332, 821, 1917.

34. Ryan to Manning (April 26, 1916), RG 70, box 237, 705, 1916; to Mitchell (June 4, 1914), Mitchell papers, box 76, file 12; W. H. Rogers to Manning (Jan. 10, 1916), RG 70, box 237, 705, 1916; and William Green, Secretary-Treasurer of UMWA, to H. M. Wilson (Sept. 15, 1913), RG 70, box 88, 1913.

35. Ryan to Manning (April 26, 1916), RG 70, box 237, 705, 1916; memorandum, Holmes to Manning (Oct. 15, 1914), RG 70, box 206, 150.3, 1916; RG 70, box 88, 1913.

36. See UMWJ, July 27, 1911, p. 7, letter from Joe King, Coulterville, Ill., July 23, describing rescue work in Alabama; Aug. 10, 1911, p. 2, another King letter; Dec. 21, 1911, p. 6, letter of J. B. McClary, Krebs, Okla., Dec. 11, 1911.

37. Editorial, UMWJ, Nov. 25, 1909, p. 4; April 28, 1910, p. 4; and Nov. 25, 1909, p. 4.

38. *Outlook* 93 (Nov. 27, 1909):639.

39. UMWJ, Jan. 20, 1910, p. 18.

40. Memorandum from Holmes, quoting a vice president and general manager of one of the large coal companies, to Mining Engineers in Charge of Cars

(Dec. 22, 1910), RG 70, box 32, 446, 1910–11. The psychological basis of the intense miner attachment to rescue programs is discussed in M. H. Ross, "Life Style of the Coal Miner: America's Original Hard Hat," *Appalachia Medicine* 3 (March 1971):7–8.

41. *Times*, Oct. 24, 1913, RG 70, box 75, 031.1, 1913.

42. Leehey to Walsh (Dec. 16, 1915), Thomas J. Walsh papers, box 225, file "Legislation-Mining."

43. Walsh to Leehey (Jan. 4, 1916), ibid.; and Walsh to Mrs. W. E. Safford of Washington, D.C. (Jan. 19, 1917), ibid., box 193.

44. Ballinger to Eugene Hale, Chairman, Senate Appropriations Committee (Feb. 25, 1911), RG 70, box 9, 031, 1910–11. See also *Estimates for Urgent Deficiency*, Bureau of Mines, S. Doc. 157, 62d Cong., 2d sess., 1912, ser. 6180, pp. 1–2; Bureau of Mines, *First Annual Report* (1911), p. 12, and *Second Annual Report* (1912), p. 78; UMWJ, June 2, 1910, p. 2.

45. Bureau of Mines, *First Annual Report* (1911), pp. 6, 8.

46. Ibid., p. 12; *Seventh Annual Report* (1917), p. 95; *Ninth Annual Report* (1919), p. 114; *Second Annual Report* (1912), p. 78; *Third Annual Report* (1913), pp. 11, 34, 37; *Fourth Annual Report* (1914), p. 30.

47. Bureau of Mines, *Fourth Annual Report* (1914), p. 34; *Third Annual Report* (1913), pp. 20–23; *First Annual Report* (1911), p. 12; RG 70, box 94, 445.1, 1913; Lane to Martin (July 18, 1914), RG 70, box 111, 031, 1914; and memorandum, Manning to Rice (July 21, 1914), RG 70, box 118, 102.4, 1914.

48. Holmes to H. M. Wilson (Feb. 16, 1912), box 60, 1912; Manning to Walsh (May 25, 1915), Walsh papers, box 193; RG 70, box 32, 445.6, 1910–11; and Dupree, *Science*, p. 159.

49. Suffern, *Conciliation and Arbitration in the Coal Industry of America* (Boston, 1915), p. 130; Fleming papers, Aug.–Dec. 1909 box.

50. U.S., Congress, House, Committee on Mines and Mining, *Bill to Change Law Establishing Bureau of Mines*, Hearings before the Committee on Mines and Mining, House of Representatives, 1912, 62d Cong., 2d sess., pp. 33, 43.

51. U.S., Congress, Senate, *Hearings on H.R. 17260*, copy in RG 70, box 48, 031, 1912, p. 6.

52. Mitchell to Foster (May 14, 1912), RG 70, box 48, 031, 1912.

53. AMC *Proceedings* 17 (1914):64.

54. 1912 House Bureau Hearings, p. 44; AMC *Proceedings* 16 (1913):97; Mining Experiment Station in Colorado, *Hearings*, 1911, p. 8.

55. *Bulletin 42 of Mining and Metallurgical Society of America* 4 (Dec. 1911): 245.

56. Typewritten copy of Jan. 20, 1912, editorial in RG 70, box 48, 031, 1912; and Channing to Martin D. Foster, RG 70, box 48, 031, 1912. For other correspondence relating to the criticism of the bill by engineers in the Mining and Metallurgical Society of America, see RG 70, box 48, 031, 1912.

57. Holmes to Callbreath (Jan. 21, 1915), RG 70, box 152, 031, 1915, and other materials in that box; to Mitchell (March 18, 1914); Secretary of the Interior to Foster (Jan. 15, 1914), in RG 70, box 110, 031, 1914; AMC *Proceedings* 17 (1914):79, 95; and W. C. Stark to Walsh (June 30, 1914), Walsh papers, box 193.

58. Holmes to W. M. Emmons, University of Minnesota, Minneapolis (Jan. 23, 1915); to President C. H. Bowman, School of Mines, Butte, Mont. (Nov.

30, 1914); to Eugene A. Smith, State Geologist, University of Alabama (Jan. 21, 1915); to Erasmus Haworth, University of Kansas (Jan. 25, 1915), RG 70, box 152, 031, 1915; Manning to DeWolf (March 24, 1915), RG 70, box 157, 131, 1915; and Holmes to William Green, Chairman, Miners' Committee on Operations of the Bureau of Mines (July 10, 1914), RG 70, box 167, 443, 1915.

59. Bureau of Mines, *Fifth Annual Report* (1915), p. 15; *Third Annual Report* (1913), pp. 38–45; *Fourth Annual Report* (1914), pp. 12, 18, 21, 49; *Sixth Annual Report* (1916), p. 4; *Seventh Annual Report* (1917), pp. 8–9.

60. Bureau of Mines, *Ninth Annual Report* (1919), pp. 13, 16; memorandum, June 18, 1918, RG 70, box 346, 031, 1918.

61. Bureau of Mines, *Seventh Annual Report* (1917), pp. 4–5; *Eighth Annual Report* (1918), pp. 7–8, 56. On the work of the bureau in explosives regulation, see RG 70, box 477, 031, 1920, and box 408, 031, 1919.

62. Van H. Manning, "The United States Bureau of Mines," *Proceedings of the Second Pan American Scientific Congress* 8 (1915–1916):485. See John B. Andrews, *Administrative Labor Legislation* (New York, 1936), pp. 21ff.

63. W. D. Ryan to Manning (July 1, 1916), RG 70, box 237, 705, 1916.

64. Bureau of Mines, *Mine-Safety Demonstration*, p. 9; "Report on Work of Pittsburg [*sic*] Station of the United States Bureau of Mines," to the Secretary of the Interior, by F. G. Cottrell, Chief Chemist, April 20, 1915, ca. 70 pages, RG 70, box 156, 101.7, 1915 (hereafter referred to as Cottrell report); and "Safety First in Mining," *Bulletin of the Pan American Union* 49 (Oct. 1919): 431. On the origins and early use of the slogan, see Dianne Bennett and William Graebner, "Safety First: Slogan and Symbol of the Industrial Safety Movement," *Journal of the Illinois State Historical Society* 68 (June 1975):243–56.

65. August F. Knoefel of the National Safety Council to Dr. Joseph Bloodgood, Baltimore (Jan. 17, 1916), Woodrow Wilson papers, Manuscript Division, Library of Congress, Washington, D.C., box 400, file 2526; Cottrell report.

66. Manning to Dr. P. O. Claxton, Commissioner of Education, Washington, D.C. (March 3, 1914), RG 70, box 107, 016.7, 1914; July 15, 1911, memorandum from the law examiner of the bureau to the Secretary of the Interior, RG 70, box 16, 1910–11; Bureau of Mines, *Second Annual Report* (1912), p. 87.

67. UMWJ, Feb. 20, 1913, pp. 6, 8.

68. Holmes to Mitchell (Dec. 5, 1910), Mitchell papers, box 15, file 63; to White (Nov. 29, 1911), RG 70, box 33, 445.6, 1910–11. See Mitchell to Holmes (Dec. 7, 1910, and Feb. 1, 1911), both in Mitchell papers, box 15, file 63, and White to Holmes (Dec. 5, 1911), RG 70, box 17, 184.2, 1910–11.

69. Bureau of Mines, *Second Annual Report* (1912), p. 20; Manning to H. A. Garfield, U.S. Fuel Administration (Feb. 15, 1918), RG 70, box 369, 444.8, 1918.

70. This, and other examples, are contained in a typewritten summary of statements made by Holmes before the Appropriations Committee at various times regarding the bureau's mine-rescue work. It is located in RG 70, box 157, 131, 1915.

71. Germany, Belgium, France, and England had private systems of rescue and training, which had been studied by bureau officials and American coal-mining executives (see Bureau of Mines, *Second Annual Report* [1912], pp. 24–25).

72. A. J. Moorshead, President and General Manager of Madison Coal Co., Illinois, to Holmes (Sept. 30, 1913), RG 70, box 94, 445.42, 1913; Rice to Manning (March 17, 1917) and memorandum, Manning to Rice (March 20, 1917), RG 70, box 220, 445, 1916.

73. Cottrell report, p. 10.

74. These tables were compiled by computer from original material in RG 70, box 221, 452, 1916. Programming was done under the direction of Robert Carr of Parkland College, Champaign, Ill.

75. From typewritten statement of Holmes in hearings, in RG 70, box 157, 131, 1915.

76. See Chapter 5.

77. Memorandum, Manning to Rice (March 20, 1917), RG 70, box 220, 445, 1916. Typewritten statement of Holmes in hearings, RG 70, box 157, 131, 1915. Typewritten cooperative agreement in RG 70, box 220, 444.7, 1916; J. J. Rutledge to Rice (Feb. 17, 1917), and Rice to Manning (March 17, 1917), RG 70, box 220, 445, 1916.

78. Manning to La Follette Coal, Iron, and Railway Co., La Follette, Tenn. (Nov. 5, 1913); Powell, *Bureau of Mines*, p. 10; and Manning, "The United States Bureau of Mines," p. 489.

79. Bureau of Mines, *Fourth Annual Report* (1914), p. 37; Powell, *Bureau of Mines*, pp. 46–47; Rice to Manning (June 16, 1915), Cottrell to Manning (June 26, 1915), and Rice to Manning (June 26, 1915), in RG 70, box 156, 101.7, 1915. Manning had a number of titles from 1910 to 1915; assistant director is used for convenience.

80. Cottrell report, p. 4.

81. RG 70, box 60, 1912.

82. D. Harrington to Rice (Aug. 4, 1916), RG 70, box 220, 444.7, 1916. A copy of the 1916 agreement is located in RG 70, box 345, 021.4, 1918. Carl A. Allen, Cooperative Engineer, to E. A. Holbrook, Acting Chief Mining Engineer (Feb. 25, 1919), RG 70, box 404, 021.4, 1919. For a list of the eighteen agreements, see "Abstracts of Cooperative Agreements Entered into between the United States Bureau of Mining and Public and Private Institutions, Persons, and Companies during the Fiscal Year 1918–1919," compiled by W. B. White. A typewritten copy is in RG 70, box 458, 700.02, 1919; see also E. A. Holbrook, "Closer Cooperation of the Colleges, United States Bureau of Mines, and Industrial Corporations in Research Investigations," AMC *Proceedings* 22 (1919): 577–78.

83. See a copy of the agreement in RG 70, box 73, 021.3, 1913; F. W. DeWolf, Illinois State Geological Survey, to Manning (July 2, 1915), RG 70, box 149, 021.3, 1915; Rice to Director (May 2, 1916), RG 70, box 188, 021.3, 1916; memorandum, Manning to Engineer in Charge (Jan. 18, 1915), RG 70, box 149, 021.3, 1915; and Bureau of Mines, *Fourth Annual Report* (1914), p. 45.

84. In 1914 there were at least four agencies besides the Bureau of Mines involved in related safety work—the Bureau of Labor Statistics, the Public Health Service, the Bureau of Standards, and the Interstate Commerce Commission. All made small contributions to coal-mining safety (see a printed report, dated Dec. 15, 1914, entitled "Activities of the Federal Government along the Lines of Safety and Sanitation," RG 70, box 191, 131.1, 1916).

85. Holmes to Willis L. Moore (Jan. 24, 1913), RG 70, box 76, 1913.

86. H. M. Wilson to D. J. Parker, Pittsburgh Station (July 19, 1916), RG 70, box 186, 016.1, 1916.

87. *Sic.* Telegram, Holmes to Page (Aug. 9, 1914), RG 70, box 191, 031.1, 1916. See H.R. 10735, 63d Cong., 2d sess., introduced by David J. Lewis of Cumberland, Md., on Dec. 16, 1913, in RG 70, box 191, 031.1, 1916.

88. Holmes to Foster (Aug. 9, 1914); telegram, Holmes to Manning (Aug. 9, 1914), RG 70, box 191, 031.1, 1916.

89. See Holmes to R. W. Raymond (Nov. 26, 1914), RG 70, box 111, 031.1, 1914. For the lobbying efforts of coal operators, see letter of Washington Coal Operators' Association to its Representatives in Congress (Feb. 24, 1916), RG 70, box 191, 031.1, 1916. For the state geologists, see W. O. Hotchkiss, Wisconsin State Geologists, to Manning (April 15, 1916), RG 70, box 189, 022.23, 1916. For the UMWA, see Manning to Holmes (Aug. 15, 1914), and John D. Ryan to Manning (Aug. 2, 1914), RG 70, box 191, 031.1, 1916. For the metal mine operators, see Holmes to Callbreath of the AMC (Aug. 9, 1914), RG 70, box 191, 031.1, 1916.

90. Rice to Engineer in Charge, Pittsburgh (Jan. 6, 1912), to Manning (Jan. 5, 1912), both in RG 70, box 47, 022.17, 1912; *Proceedings of the National Fire Protection Association* (NFPA), Fifteenth Annual Meeting, New York, May 23–25, 1911 (n.p., n.d.), pp. 286, 288.

91. NFPA *Proceedings*, 1912, pp. 114, 116; NFPA *Newsletter*, Boston, Nos. 2, 4, 5, 6–16, 17, 18–30, 33–39, 41–42 (1917–1920); Manning to H. M. Wilson (Oct. 11, 1913), RG 70, box 74, 022.421, 1913.

92. Holmes to H. M. Chance (June 29, 1914), RG 70, box 150, 022.10, 1915.

93. Acting Director Manning to Saunders (Aug. 23, 1915), RG 70, box 189, 022.10, 1916. Holmes to Charles F. Rand (Nov. 24, 1913), and Rand to Holmes (Nov. 28, 1913), RG 70, box 73, 022.07, 1913; Bureau to AIME (June 20, 1914), RG 70, box 150, 022.10, 1915; Holmes to Chance (June 29, 1914), to Benjamin B. Thayer, President of AIME (June 19, 1914), and B. B. Thayer to Holmes (June 24, 1914), all in RG 70, box 150, 022.10, 1915; and Holmes to Chance (July 21, 1914), RG 70, box 189, 022.10, 1916.

94. Memorandum, Rice to Manning (March 20, 1916), RG 70, box 189, 022.10, 1916.

95. Holmes to C. C. Mailloux, AIEE President (April 5, 1914), Manning to H. H. Clark (April 14, 1914), Wilfred Sykes, Chairman of Committee on Electricity in Mining, to Bureau (April 14, 1914), Holmes to Sykes (April 15, 1914), all in RG 70, box 189, 022.08, 1916; Manning to H. W. Buck, AIEE President (March 3, 1917), and Buck to Director (March 27, 1917), RG 70, box 243, 022.08, 1917.

96. Telegram, Holmes to H. M. Wilson at Bureau of Mines (July 23, 1914), Wilson to Holmes (Sept. 18, 1914), RG 70, box 109, 022.59, 1914; and George S. Pope, Acting Chief Clerk, to Edgar Marburg, Secretary-Treasurer of American Society for Testing Materials (Sept. 2, 1913), RG 70, box 109, 022.13, 1914.

97. H. M. Wilson to Holmes (Jan. 19, 1915), and O. P. Hood to Director (March 27, 1916), RG 70, box 151, 022.59, 1915.

98. The commission issued certificates for rescue and first-aid work. Another object was to secure the establishment of rescue stations in the state of Illinois (RG 70, box 149, 021.2, 1915). Holmes was one of the three public representatives on the Illinois Mining Investigation Commission, an organization of pub-

lic, miners, and operators designed to prepare legislation for submission to the legislature. It was founded about the same time—1910 (see RG 70, box 8, 1910–11, and RG 70, box 150, 022.47, 1915).

99. Director of Bureau to A. F. Knoefel, Terre Haute, Ind. (May 5, 1917), RG 70, box 332, 021, 1917; Bureau of Mines, *Third Annual Report* (1913), p. 67; RG 70, box 105, 013.4, 1914; Bureau of Mines, *Fourth Annual Report* (1914), p. 25; and RG 70, box 47, 022.26, 1912, particularly Holmes to W. B. Cannon, Harvard University (March 6, 1912) and Holmes to Abraham Jacobi, AMA President (April 2, 1912).

100. American Mine Safety Association (AMSA), *First Annual Transactions, Constitution and By-Laws*, held at Pittsburgh, Pennsylvania, January 1914 (n.p., n.d.), pp. 4–6, 9, 23–24, 46, 48, 52, 53, 59, 61; AMSA, *Second Annual Transactions*, 1915, pp. 5, 60; Bureau of Mines, *Third Annual Report* (1913), p. 68; and H. M. Wilson, "Progress of Safety in Mining," National Safety Council (NSC) *Proceedings*, Mining Section, 5 (1916):868.

101. Wilson, "Progress of Safety in Mining," p. 869; NSC *Proceedings* 4 (1915): 13; AMSA, *First Annual Transactions*, 1914, p. 48; and Bureau of Mines, *Third Annual Report* (1913), p. 68.

102. Undated memorandum from W. D. Ryan, RG 70, box 244, 1917; D. J. Parker to Chief Mining Engineer (June 14, 1917), RG 70, box 244, 022.72, 1917; series of letters dated Dec. 1913 to John Mitchell, John P. White, and William Green from Lane, RG 70, box 117, 102.1, 1914; Holmes to Thomas Kennedy (Jan. 24, 1914), RG 70, box 117, 102.1, 1914; Van Amberg Bittner papers, box 1, file "Correspondence 1919," in West Virginia University Library, Morgantown; and UMWA, International Executive Board, *Minutes*, meeting of Feb. 4, 1912, p. 176 (in MS).

103. Holmes to R. W. Raymond (Nov. 26, 1914), RG 70, box 111, 031.1, 1914.

104. Dupree, *Science*, pp. 159–60.

CHAPTER 3

1. *The Code of West Virginia*, 1906, National Reporter System (St. Paul, Minn., 1906), sec. 410. This description is also compiled from *Hogg's West Virginia Code, Annotated*, ed. Charles E. Hogg, 3 vols. (St. Paul, Minn., 1914), vol. 1; and Glenn F. Massay, "Legislators, Lobbyists, and Loopholes: Coal Mining Legislation in West Virginia, 1875–1901," *West Virginia History* 32 (April 1971):135–70.

2. Editorial, UMWJ, March 15, 1900, p. 4.

3. West Virginia, *Journal of the Senate of the State of West Virginia for the Twenty-fifth Regular Session, Commencing January 9, 1901* (Charleston, 1901), p. 21 (hereafter referred to as *W. Va. Senate Journal*).

4. Ibid., pp. 120, 204; West Virginia, *Journal of the House of Delegates of the State of West Virginia for the Twenty-sixth Regular Session, Commencing January 14, 1903* (Charleston, 1903), pp. 23–24, Appendix B, p. 368 (hereafter referred to as *W. Va. House Journal*); *1905*, p. 626.

5. West Virginia, *Sixteenth, Seventeenth and Eighteenth Annual Reports, Coal Mines in the State of West Virginia, U.S.A., For the Years Ending June 30, 1898,*

1899, 1900, compiled by James W. Paul, Chief Mine Inspector (Charleston, n.d.), *1900*, pp. 373, 374–78 (hereafter referred to as *W. Va. Inspectors' Report*, followed by date of fiscal year); *Coal* 3 (March 3, 1906):27.

6. *W. Va. Inspectors' Report, 1902*, pp. 150, 4.

7. Ibid., p. 3; ibid., *1903*, pp. 7, 8, 159.

8. Ibid., *1905*, pp. 2, 10; ibid., *1906*, pp. 13–14; *Coal Trade Bulletin* 12 (May 15, 1905):36.

9. *Hogg's West Virginia Code*, vol. 1, secs. 469–518.

10. *W. Va. House Journal, 1907*, pp. 653, 815–17; *W. Va. Senate Journal, 1907*, pp. 581–83, 585–86, 606, 774–76; *W. Va. Inspectors' Report, 1907*, pp. x, ix. The 1907 legislation was passed 69-2-15 in the House, 19-3-8 in the Senate.

11. *W. Va. House Journal, 1907*, pp. 406, 407, 423–24, 438.

12. Ibid., *1908*, pp. 22, 25, 27.

13. *W. Va. Hearings, 1909*, p. 821; *W. Va. House Journal, 1908 extra session*, pp. 36–39; *W. Va. Inspectors' Report, 1908*, pp. 448, 466–68, 491–92; and *1909*, pp. 425, 513.

14. *W. Va. House Journal, 1908*, pp. 323, 326–27, 359–60.

15. Ibid., pp. 284, 688; *W. Va. Senate Journal, 1908*, pp. 256–57; *Coal Trade Bulletin* 18 (March 2, 1908):26; 20 (March 1, 1909):49; 19 (Oct. 15, 1908):30; and *Coal* 7 (March 19, 1908):18.

16. West Virginia, *Acts of the Legislature of West Virginia, Regular and Extraordinary Sessions, 1915* (Charleston, 1915).

17. *W. Va. Inspectors' Report, 1910*, pp. 10, 16–18; *1912*, pp. 11, 196; and *W. Va. Senate Journal, 1909*, Appendix A, p. 49.

18. Compiled from *The Annotated Revised Statutes of the State of Ohio, Including All Laws of a General Nature in Force January, 1902*, ed. Clement Bates, 3d ed., 3 vols. (Cincinnati, 1900), vol. 1.

19. *Thirty-second Annual Report of the Chief Inspector of Mines to the Governor of the State of Ohio, for the Year Ending December 31, 1906* (Columbus, 1907), p. 20; *1905*, pp. 9, 352 (hereafter referred to as *Ohio Inspectors' Report*); *The Journal of the House of Representatives of the State of Ohio, for the Regular Session of the Seventy-fifth General Assembly, Commencing Monday, January 6, 1902* (Cincinnati, 1902), 95:1500; *1904*, 96:472; *1906*, 97:509 (hereafter referred to as *Ohio House Journal*); *The Journal of the Senate of the Seventy-sixth General Assembly of the State of Ohio for the Regular Session Commencing Monday, January 4, and Ending Monday, April 25, 1904* (Springfield, 1904), 96: 477; *1906*, 97:156 (hereafter referred to as *Ohio Senate Journal*).

20. *The State of Ohio General and Local Acts Passed and Joint Resolutions Adopted by the 76th General Assembly, 1906* (Springfield, 1906), 98:95, 259 (hereafter referred to as *Ohio Laws*); *Ohio Inspectors' Report, 1906*, p. 9; *Ohio House Journal, 1906*, pp. 832–33.

21. *Ohio Inspectors' Report, 1907*, p. 114, also pp. 23–26, 330; *1908*, pp. 7, 45–47; *Ohio House Journal, 1908*, pp. 244–45, 777–78; *Ohio Senate Journal, 1908*, pp. 379, 384; *Ohio Laws, 1908* (Springfield, 1908), 99:106–14.

22. *Ohio House Journal, 1908*, p. 889; *Ohio Senate Journal, 1908*, p. 480, and Appendix, p. 29; *Ohio Inspectors' Report, 1908*, pp. 6–7; *1909*, pp. 9, 15, 16, 32; and *1910*, pp. 59–60.

23. *Ohio Laws, 1910; Ohio Inspectors' Report, 1910*, pp. 69–70; *Ohio Senate Journal, 1910*, p. 204; and *Ohio House Journal, 1910*, pp. 483–84.

24. *Ohio Inspectors' Report, 1910*, pp. 346–47; and *1912*, p. 437.

25. "Report of the Ohio Coal Mining Commission to the Governor," Commission established under House Joint Resolution 38, 80th General Assembly, Dec. 17, 1913, in *Ohio Senate Journal, 1914*, pp. 328–33.

26. Editorial, *Colliery Engineer* 34 (Feb. 19, 1914):397; *Coal and Coke Operator and Fuel Magazine* 18 (Jan. 15, 1914):38–39.

27. *Ohio Inspectors' Report, 1914*, pp. 10–13, 37–42, 67; *Ohio House Journal, 1914*, p. 36. In 1919 two other fairly important pieces of legislation were passed. One created five rescue stations in the coal areas; the other amended the electricity laws.

28. Since accurate and relatively detailed descriptions of the process and content of Illinois and Pennsylvania mine safety law are available in print, in Beckner, *Labor Legislation*, and Trachtenberg, *Protection of Coal Miners*, the following discussions are limited to those elements in the legislative history of particular importance and relevance.

29. *Mines and Minerals* 29 (Aug. 1908):18–19; Beckner, *Labor Legislation*, pp. 354–67; *Nineteenth Annual Coal Report of the Illinois Bureau of Labor Statistics, 1900, for the Year Ending October 1, 1900*, compiled by David Ross, Secretary, Illinois Bureau of Labor Statistics (Springfield, 1901), pp. 2–3; *1904*, pp. 2–3; *1906*, pp. xiv–xvi; *1908*, pp. 3–4; *1909*, p. 150 (hereafter referred to as *Illinois Coal Report*).

30. *Illinois Coal Report, 1910*, p. 4; *Proceedings of the Twenty-first Annual Convention of the United Mine Workers of America, District 12, Held in the City of Peoria, Illinois, February 15 to March 5, Inclusive, 1910* (n.p., n.d.), pp. 15, 125, 131 (hereafter referred to as UMWA, *District 12 Proceedings*); Eugene Staley, *History of the Illinois State Federation of Labor*, Social Science Studies 15 (Chicago, 1930):267–68.

31. *Joint Conference of the Illinois Coal Operators Association, the Coal Operators Association of the Fifth and Ninth Districts of Illinois, the Central Illinois Coal Operators Association and the* UMWA *District No. 12, Peoria, Illinois, April 2 to May 8, 1914* (n.p., n.d.), pp. 8, 299, 302, 314, 317 (hereafter referred to as *Joint Conference, District 12*); and Beckner, *Labor Legislation*, pp. 296–309, 364–65.

32. Editorial, *Coal* 5 (May 23, 1907):18.

33. *Black Diamond* 28 (Feb. 15, 1902):232; 32 (Jan. 23, 1904):186; *Mines and Minerals* 27 (Sept. 1906):67; *Coal* 6 (Dec. 5, 1907):20; and *Coal and Timber* 1 (Feb. 1905):10.

34. *Report of the Bureau of Mines of the Department of Internal Affairs of Pennsylvania, 1901* (Wm. Stanley Ray, State Printer of Pa., 1902), pp. 3, 10, 16; *1903*, pt. 2, pp. v, vi, xviii; *1904*, pt. 2, p. liv; and *1906*, pt. 2, pp. 24, 28 (hereafter referred to as *Pa. Mines Report*); *Black Diamond* 32 (Jan. 23, 1904):186.

35. *Pa. Mines Report, 1907*, pt. 2, pp. xiii, xii; *1905*, pt. 2, p. xvii.

36. Ibid., *1909*, pt. 2, pp. liii, liv, lv; *1908*, pt. 2, pp. 44, 46; *Coal Trade Bulletin* 20 (March 1, 1909):22–23, 47–48; (April 15, 1909):24, 48; and *Coal* 7 (Jan. 23, 1908):23.

37. *Pa. Mines Report, 1909*, pt. 2, p. liv; *Coal Trade Bulletin* 20 (April 15, 1909):26.

38. *Pa. Mines Report, 1908*, pt. 2, pp. 46–47; *1909*, pt. 2, pp. xlviii, xlix, li, lviii, lx.

39. *Mines and Minerals* 32 (April 1912):527; *Pa. Mines Report, 1913*, pt. 2, p. 13; *1911*, pt. 2, pp. v–vi; *Coal Trade Bulletin* 25 (June 15, 1911):21; *Coal Age* 1 (Jan. 6, 1912):413; UMWA, *Minutes of the 22nd Annual Convention of District No. 2, Dubois, Pennsylvania, March 21–30, 1911*, 2 vols. (n.p., n.d.), 1:26; Commonwealth of Pennsylvania, *Legislative Journal for the Session of 1911* (Harrisburg, 1912), pp. 3521–23 (hereafter referred to as *Pa. Legislative Journal*).

40. *Pa. Mines Report, 1908*, pt. 2, p. 42.

41. *Pa. Legislative Journal, 1911*, p. 3351; *Pa. Mines Report, 1908*, pt. 2, pp. 26–27, 42; *1907*, pt. 2, p. xxxiii; and *1909*, pt. 2, p. iv.

42. *Pa. Legislative Journal, 1911*, p. 3351.

43. Ibid.; *Pa. Mines Report, 1915*, pt. 2, p. 8; and *Digest of Pennsylvania Statute Law, 1920 (complete), based on Pepper and Lewis' Digest Laws of Pennsylvania* (St. Paul, Minn., 1921), pp. 1475–520.

44. This information was obtained from Pennsylvania, *Laws of the General Assembly of the Commonwealth of Pennsylvania, Sessions 1899–1917 (odd years)*, (Harrisburg, 1899–1917); *Acts of the Legislature of West Virginia, Sessions 1899–1919 (odd years)*, (Charleston, 1899–1919); State of Ohio, *General and Local Acts Passed and Joint Resolutions Adopted by the General Assembly at Regular Sessions Begun and Held in the City of Columbus, Sessions 1900–1920* (Columbus and Springfield, 1900–1920); and U.S. Geological Survey, *Mineral Resources of the United States, 1900–1920* (Washington, D.C., 1901–1921). For Pennsylvania, because the appropriations for anthracite and bituminous mining were not separated, the percentage of expenditures applicable to bituminous mining was calculated as a function of tonnage. Actually, bituminous tonnage was a relatively constant percentage of total tonnage in the state in this period (58–64%). All expenditures were corrected for inflation and deflation, using a wholesale price index for all commodities (Series E 13–24 in U.S., Bureau of the Census, *Historical Statistics*, p. 116). Total expenditures in each year were divided by total tonnage in that year by state. Some of the large fluctuations are the result of tonnage increases or decreases rather than expenditures on mine safety.

45. *W. Va. Inspectors' Report, 1900*, p. 353; *1902*, pp. 5, 215; *1903*, pp. 7, 220; *1906*, p. 14; *1908*, p. 448; and *1914*, sec. 3, p. 211.

46. *Annual Reports of the Inspector of Mines of the State of Kentucky, for 1903 and 1904, prepared by C. J. Norwood, Chief Inspector* (Louisville, 1905), pp. 4, 11; *1907*, p. 1; *1908*, p. 44; *1914*, p. 4 (hereafter referred to as *Kentucky Inspectors' Report*); and *Coal Age* 5 (April 25, 1914):676–77.

47. Table compiled from *Ohio Inspectors' Report, 1898*, p. 55, and *1899*, pp. 32–33; also *1906*, pp. 10–11.

48. Trachtenberg, *Protection of Coal Miners*, p. 177; *Pa. Mines Report*, 1900, p. 473; *Coal Trade Bulletin* 22 (Jan. 15, 1910):44–46; and UMWJ, Feb. 18, 1915, p. 29.

49. *Pa. Mines Report, 1907*, pt. 2, p. xxx; *1906*, pt. 2, p. 23; and *1904*, pt. 2, p. xi.

50. Figures for numbers of inspectors were taken from Beckner, *Labor Legislation*, Trachtenberg, *Protection of Coal Miners*, and the inspectors' reports; tonnage figures are largely from *Mineral Resources*.

51. "Biennial Message of Governor John J. Cornwell," in *West Virginia Public Documents, 1917–1918*, 3 vols. (n.p., n.d.), pt. 1, p. 23; *W. Va. Inspectors' Report, 1900*, p. 353; *1912*, sec. 3, p. 291; *W. Va. House Journal, 1905*, p. 58; and *W. Va. Senate Journal, 1915*, pp. 262–63, 267; *Ohio Inspectors' Report, 1905*, p. 10; *1906*, pp. 10–12; *Ohio House Journal, 1908*, p. 964; *Kentucky Inspectors' Report, 1908*, pp. 4–5.

52. *Illinois Coal Report*, 1903, pp. 134–40; *1918*, p. iii; *Ohio Inspectors' Report, 1905*, pp. 14–15; *W. Va. House Journal, 1908*, p. 40; *W. Va. Senate Journal, 1909*, Appendix A, pp. 47–48; UMWJ, Oct. 10, 1912, pp. 1, 4; July 29, 1909, p. 4; and *Coal and Coke Operator* 9 (July 15, 1909):19.

53. May 17, 1900, p. 4.

54. *W. Va. Inspectors' Report, 1908*, p. 424; *Ohio Inspectors' Report, 1911*, p. 68; UMWJ, May 17, 1900, p. 4; and Roy, *Coal Miners*, pp. 146–47.

55. UMWA, *District 12 Proceedings, 1910*, p. 95; Charles P. McGregor to Van Bittner (April 22, 1915), and Bittner to McGregor (April 22, 1915), in Van Amberg Bittner papers, box 1, file "Correspondence 1914"; Chris Evans, *History of the United Mine Workers of America*, 2 vols. (Indianapolis, n.d.), 2:707; *Coal Trade Bulletin* 24 (Jan. 2, 1911):34; *W. Va. Inspectors' Report, 1909*, pp. 449–50.

56. *Report of Industrial Commission on Capital and Labor*, p. 161.

57. UMWJ, Dec. 21, 1911, p. 3; Sept. 30, 1915, p. 5; Nov. 7, 1912, p. 6; March 30, 1911, p. 7; UMWA, *District 12 Proceedings, 1910*, p. 95; *1913*, p. 88; *1918*, pp. 127, 140–41; UMWA *Proceedings*, 1910, pp. 577–78; *Mines and Minerals* 31 (May 1911):611; 32 (Oct. 1911):130; (June 1912):634; and letter, Van Bittner to UMWJ (Dec. 12, 1907), in Bittner papers, box 10, file "1910–1918."

58. Frank E. Parson, Inspector 2d District, to Governor Elect William E. Glasscock, and John A. Springer, Inspector 4th District, to Glasscock (Feb. 18, 1909), in William E. Glasscock papers, West Virginia University Library, Morgantown; *W. Va. Senate Journal, 1901*, p. 204; *1917*, pp. 240–41; *W. Va. House Journal, 1905*, p. 307; *W. Va. Public Documents, 1917–1918*, p. 24. A primitive 1883 Ohio competency statute was apparently not invoked. Under it, competency examinations were initiated only on petition.

59. *W. Va. Inspectors' Report, 1915*, p. 268.

60. Roan to Lewis Minute (Oct. 16, 1914), Ohio, Department of Industrial Relations, Division of Mines, State Letters, Ohio State Historical Society, Columbus, vol. "May 19 to Oct. 19, 1914" (hereafter referred to as Roan papers); UMWA, *District 12 Proceedings, 1910*, p. 92; *Kentucky Inspectors' Report, 1901 and 1902*, p. 4; *Ohio Inspectors' Report, 1906*, p. 351.

61. J. M. Roan to Forsythe Coal Co., Cambridge, Ohio (Feb. 16, 1915), Roan papers, vol. "Oct. 19, 1914–Feb. 17, 1915"; to Earl R. Lewis (Nov. 27, 1914), Roan papers, vol. "Oct. 19–Feb. 17, 1914."

62. Roan to Wolf Run Mining Co., Cleveland (Dec. 14, 1914), Roan papers, vol. "Oct. 19, 1914–Feb. 17, 1915"; to J. A. Collins (Nov. 16, 1914), ibid.

63. Roan to Deputy Mine Inspectors, State of Ohio (May 29, 1914), in Roan papers, vol. "May 19 to Oct. 19, 1914."

64. UMWJ, Oct. 11, 1906, p. 4.

65. UMWJ, June 5, 1902, p. 4.

66. *Coal and Coke Operator* 8 (April 1, 1909):234; UMWA, *District 12 Pro-*

ceedings, 1910, p. 95; *Report of Industrial Commission on Capital and Labor*, p. 150; UMWJ, Feb. 11, 1915, p. 9; Evans, *History of the UMWA*, 2:706; John Brophy, *A Miner's Life* (Madison, Wis., 1964), p. 39.

67. *W. Va. Inspectors' Report, 1902*, p. 148; *1909*, p. 308; *1910*, sec. 3, p. 217.

68. *W. Va. Hearings, 1909*, p. 824; *W. Va. Inspectors' Report, 1908*, p. 491; *1907*, p. 321; *Illinois Coal Report, 1904*, pp. 231–32; *1906*, p. xiv; *Kentucky Inspectors' Report, 1914*, p. 12; *Ohio Inspectors' Report, 1909*, p. 402; *Coal Trade Bulletin* 12 (Jan. 16, 1905):27; and *Joint Conference, District 12, 1902*, pp. 188–89.

69. Roan to Robert Kidd (June 10, 1914), Roan papers, vol. "May 19 to Oct. 19, 1914."

70. *Pa. Mines Report, 1904*, pt. 2, p. ix; *1907*, pt. 2, p. xii; *1911*, pt. 2, pp. v–vi. The subject of miner carelessness is treated at greater length in Chapter 4.

71. *W. Va. House Journal, 1908*, p. 26.

72. *W. Va. Hearings, 1909*, p. 824; *Kentucky Inspectors' Report, 1901 and 1902*, pp. 41–43; *W. Va. Inspectors' Report, 1908*, p. 309; *1905*, p. 204; *Ohio Inspectors' Report, 1899*, p. 61.

73. *W. Va. Inspectors' Report, 1906*, p. 254; *1905*, pp. 203, 242; *Ohio Inspectors' Report, 1906*, p. 348; *1905*, p. 270.

74. *W. Va. Inspectors' Report, 1900*, pp. 361–71.

75. *Ohio Inspectors' Report, 1908*, p. 242; *Black Diamond* 28 (March 1, 1902):315; *Joint Conference, District 12, 1902*, p. 77; Powers Hapgood, *In Non-Union Mines: The Diary of a Coal Digger in Central Pennsylvania, August–September, 1921* (New York, n.d.), pp. 38–42; *W. Va. Inspectors' Report, 1906*, p. 399; *1907*, p. 443; *1908*, p. 510; and *1902*, p. 4.

76. *W. Va. Inspectors' Report, 1908*, pp. 509–10; *Ohio Inspectors' Report, 1906*, p. 236; *1910*, p. 240; UMWJ, Feb. 11, 1915, p. 9; and UMWA, International Executive Board, *Minutes*, meeting of Feb. 4, 1908, pp. 90–91 (in MS).

77. *Ohio Inspectors' Report, 1909*, pp. 45–46.

78. *Kentucky Inspectors' Report, 1903–4*, p. 30; *Ohio Inspectors' Report, 1898*, p. 175; *1909*, p. 69; *Illinois Coal Report, 1905*, p. 268; *W. Va. Inspectors' Report, 1906*, pp. 229, 316; *1907*, p. 388; *1908*, pp. xvi, 421. This picture is developed from inspectors' reports and might be considerably modified by an examination of court records.

79. *W. Va. Senate Journal, 1909*, Appendix A, p. 48.

80. UMWJ, April 21, 1904, p. 4; April 28, 1904, p. 4; Feb. 22, 1906, p. 4; March 1, 1906, p. 1; Jan. 8, 1914, p. 2; *Coal Trade Bulletin* 12 (Jan. 16, 1905): 27; *W. Va. Inspectors' Report, 1900*, p. 307; *1906*, p. 5.

81. *Pa. Mines Report, 1911*, pt. 2, pp. xi–xviii; *1904*, pt. 2, p. 244; *Ohio Inspectors' Report, 1911*, p. 79; *1906*, p. 20; *Kentucky Inspectors' Report, 1903–4*, p. 30.

82. This description is compiled from state inspection reports, 1900–1920.

83. (Sept. 16, 1914), Roan papers, vol. "May 19 to Oct. 19, 1914."

84. *W. Va. Inspectors' Report, 1908*, p. xvi.

85. Ibid., *1907*, p. 443.

86. *Mines and Minerals* 27 (March 1907):354–55; *Black Diamond* 31 (Oct. 31, 1903):890; *Coal Trade Bulletin* 17 (June 1, 1907):24.

87. *Report of the Industrial Commission on Labor Legislation*, 5:4; U.S., Committee on Public Information, *Bulletin* 3, no. 52 (Jan. 29, 1919):9; U.S.,

Congress, House, Committee on Labor, *Child Labor Bill,* Hearings before the Committee on Labor, House of Representatives, on H.R. 12292, 63d Cong., 2d sess., 1914, p. 5; UMWJ, Jan. 6, 1910, p. 2; U.S., Congress, Senate, Committee on Public Health and National Quarantine, *Proposed Department of Public Health,* Hearings before the Committee on Public Health and National Quarantine, Senate, on S. 6049, 1910, p. 95; *Bulletin of the American Institute of Mining Engineers* 74 (Feb. 1913):319–29; *Leslie's Weekly* 104 (March 7, 1907):216.

88. *Report of the Uniform Mining Laws Conference, Chicago, Illinois, November 13, 14, 15, 1916* (Springfield, n.d.), p. 47 (hereafter referred to as *Uniform Mining Laws Conference, 1916*).

89. *Coal Trade Bulletin* 11 (July 15, 1904):30; *Uniform Mining Laws Conference, 1916,* pp. 29–30, 49; *Coal and Coke Operator and Fuel Magazine* 18 (June 25, 1914):395; and *Colliery Engineer* 35 (Aug. 1914):37.

90. MIIA *Proceedings,* 1920, p. 35.

91. To Manning (Nov. 4, 1916), RG 70, box 191, 033, 1916; *Uniform Mining Laws Conference, 1916,* p. 23; *Mines and Minerals* 33 (Feb. 1913):345–46.

92. AMC *Proceedings* 14 (1911):41. Workmen's compensation was one of the most highly developed subjects of uniformity campaigns.

93. Ibid., 16 (1913):62; *Coal and Coke Operator* 21 (Oct. 1917):108–9; *Black Diamond* 30 (Jan. 31, 1903):200.

94. UMWJ, Feb. 26, 1914, p. 4; June 18, 1914, p. 4; *Uniform Mining Laws Conference, 1916,* pp. 9–11; AMC *Proceedings* 19 (1916):600.

95. *Coal and Coke Operator and Fuel Magazine* 18 (June 25, 1914):395; *Uniform Mining Laws Conference, 1916,* pp. 5, 29; MIIA *Proceedings,* 1914, pp. 27–28; AMC *Proceedings* 19 (1916):585–97; *Coal Age* 6 (Nov. 14, 1914):795.

96. Charles McCarthy, "Need of Information about Interstate Competition," *Survey* 22 (Aug. 21, 1909):697; manuscript copy of address given by Andrews sometime in 1921 or 1922, in American Association for Labor Legislation (AALL) papers, Labor and Industrial Relations Institute, Cornell University, Ithaca, N.Y., Record Group A–1, Box 69, p. 11.

97. Respectively, John D. Works (see RG 70, box 75, 031, 1913), and George A. Pearre (see *Congressional Record,* 57th Cong., 1st sess., 1902, 35, pt. 2:1129); the Moorshead statement is in AMC *Proceedings* 16 (1913):63.

98. UMWJ, Dec. 28, 1911, p. 6; Dec. 19, 1907, p. 2; and AMC *Proceedings* 19 (1916):600.

99. NSC *Proceedings* 1 (1912):10–11; Manning to H. M. Wilson (Oct. 11, 1913), RG 70, box 74, 022.42, 1913; AMC *Proceedings* 11 (1908):95–96.

100. 105 (Sept. 13, 1913):57.

101. UMWA *Proceedings,* 1914, pp. 326–27; *Uniform Mining Laws Conference, 1916,* p. 11.

102. UMWJ, Aug. 6, 1914, p. 2 (clipping from *British Federationist*); *Uniform Mining Laws Conference, 1916,* p. 9.

103. American Bar Association, *Journal* 2 (1916):666.

104. Raymond Zillmer, "The Commissioners on Uniform State Laws," *American Law Review* 47 (Nov.–Dec. 1913):864.

105. UMWA *Proceedings,* 1906, p. 48; Mitchell to William Taft (Nov. 27, 1909), Mitchell papers, box 134, file 191.

106. *Coal Trade Bulletin* 20 (Dec. 15, 1908):40.

107. *Mines and Minerals* 28 (March 1908):371.

108. This plan is mentioned in most of the annual reports issued by the Bureau of Mines. See Bureau of Mines, *Fifth Annual Report* (1915), p. 7; also NSC *Proceedings*, Mining Section, 4 (1915):227; Holmes to Governor John K. Tener, Pa. (March 19, 1913), RG 70, box 75, 033, 1913.

109. MIIA *Proceedings*, 1910, pp. 22–23; 1909, p. 21; 1912, pp. 15–16; 1915, pp. 78–82; NSC *Proceedings*, Mining Section, 5 (1916):889, 892; 6 (1917):1401; *Coal Age* 9 (March 4, 1916):422; RG 70, box 40, 1910–11; box 55, 1912; box 104, 710, 1913; Bureau of Mines, *Sixth Annual Report* (1916), pp. 42–43.

110. AMC *Proceedings* 12 (1909):75–76; (1908):60; (1910):136–49; H. M. Wilson to W. H. Cameron, National Safety Council (Jan. 6, 1915), RG 70, box 163, 433, 1915; S. W. Stratton, Director, Bureau of Standards, to Holmes (Oct. 14, 1914), and Holmes to Stratton (Sept. 30, 1914), RG 70, box 107, 018.11, 1914; TAIME 48 (1914):243–46; Bureau of Mines, *Fifth Annual Report* (1915), p. 8; MIIA *Proceedings*, 1920, pp. 34–37.

111. MIIA *Proceedings*, 1920, pp. 14–16; 1908, pp. 9–10, 18; 1909, pp. 7–8, 13, 20; 1910, pp. 127–28, 152–53; AMC *Proceedings* 11 (1908):59.

112. MIIA *Proceedings*, 1915, pp. 34–35, 52–54, 66; 1919, pp. 18, 21–22, 33. No meetings of the institute were held in 1917 and 1918.

113. Ibid., pp. 43–44.

114. *Uniform Mining Laws Conference, 1916*, pp. 17–18, 45–46; MIIA *Proceedings*, 1915, p. 55; 1916, p. 24.

115. AMC *Proceedings* 11 (1908):103; 19 (1916):572.

116. Ibid., 16 (1913):64; MIIA *Proceedings*, 1910, pp. 13, 72.

117. From a typewritten report by Governor Frank O. Lowden to the 50th General Assembly, 1917, quoted in Louis Bloch, *Labor Agreements in Coal Mines* (New York, 1931), p. 278; Beckner, *Labor Legislation*, p. 304; and AMC *Proceedings* 14 (1911):12.

118. AMC *Proceedings* 19 (1916):146; H. H. Stoek to Manning (Sept. 26, 1916), and Manning to Stoek (Oct. 2, 1916), in RG 70, box 191, 033, 1916.

119. AMC *Proceedings* 16 (1913):62; *Uniform Mining Laws Conference, 1916*, Addendum, p. 70; AMC *Proceedings* 19 (1916):154, 158; and Beckner, *Labor Legislation*, p. 305.

120. MIIA *Proceedings*, 1913, p. 16.

CHAPTER 4

1. 12 (May 1, 1905):7; 18 (Dec. 16, 1907):24; 19 (Sept. 15, 1908):22; and 22 (Dec. 1, 1909):42. See also *Coal* 5 (Jan. 10, 1907):24; *Illinois Coal Report, 1919*, p. 1; *W. Va. House Journal, 1905*, p. 57; *Black Diamond* 31 (Oct. 3, 1903): 682.

2. Eastman, *Work-Accidents*, p. 86.

3. Ibid., pp. 84–85.

4. Industrial Workers of the World, *Coal-Mine Workers and Their Industry* (n.p., n.d.), p. 43.

5. To P. A. Grady, Mine Inspector, District 12, West Virginia (March 25, 1912), RG 70, box 62, 441, 1912; *Coal Age* 10 (Dec. 9, 1916):974–75.

6. UMWJ, Aug. 9, 1917, p. 4; and Duncan McDonald, unpublished auto-

biography, in Duncan McDonald Collection, Illinois State Historical Society, Springfield, box 2, folder 13, pp. 18–19.

7. *Pa. Mines Report, 1900*, p. xxx; *1901*, p. 509; *1905*, pp. xxvi, xxi; *1907*, p. xxxiii; *1909*, p. iv; *1912*, p. 57; *Ohio Inspectors' Report, 1906*, p. 10; *Illinois Coal Report, 1910*, p. 351; UMWJ, April 4, 1915, p. 17, and Jan. 30, 1913, p. 3.

8. *Mines and Minerals* 30 (Dec. 1909):291; *Coal* 7 (Dec. 3, 1908):18–19; *Coal and Coke Operator* 9 (Sept. 9, 1909):152.

9. *Mines and Minerals* 28 (July 1908):562; *W. Va. Inspectors' Report, 1909*, p. 309.

10. Homer Lawrence Morris, *The Plight of the Bituminous Coal Miner* (Philadelphia, 1934), pp. 62–69 (quotation on p. 67); Carter Goodrich, *The Miner's Freedom* (Boston, 1925); Beckner, *Labor Legislation*, p. 289.

11. *Colliery Engineer* 33 (July 1913):722.

12. Goodrich, *Miner's Freedom*, pp. 36–37; *Black Diamond* 33 (Sept. 24, 1904):694; *Coal Trade Bulletin* 22 (May 16, 1912):31; *W. Va. Inspectors' Report, 1900*, p. 312; *1908*, p. 538; *1910*, pp. 16–17; J. M. Roan to Robert Kidd (June 10, 1914), in Roan papers, vol. "May 19 to Oct. 19, 1914."

13. To Holmes (Oct. 6, 1910), RG 70, box 35, 474, 1910–11.

14. Goodrich, *Miner's Freedom*, p. 18, quoted from H. A. Haring, "Three Classes of Labor to Avoid," *Industrial Management* (Dec. 1921), pp. 370 ff.

15. Goodrich, *Miner's Freedom*, p. 57.

16. Brophy, *Miner's Life*, p. 41; also Eastman, *Work-Accidents*, p. 38.

17. Anna Rochester, *Labor and Coal*, Labor and Industry Series (New York, 1931), p. 149; *Coal Trade Bulletin* 22 (Dec. 15, 1909):53; (Aug. 1, 1908):51; *Coal Age* 2 (Sept. 21, 1912):406; TAIME 10 (May 1881–Feb. 1882):68, 70; and Groves report.

18. Frank Julian Warne, "The Effect of Unionism upon the Mine Worker," *Annals of the American Academy of Political and Social Science* 21 (Jan. 1903): 34.

19. *Outlook* 93 (Sept. 25, 1909):196.

20. Letter signed A. M. Iner in *Coal Age* 9 (May 27, 1916):940; *W. Va. Inspectors' Report, 1907*, p. ix; and Jerold S. Auerbach, "Progressives at Sea: The LaFollette Act of 1915," *Labor History* 2 (Fall 1961):360.

21. Bureau of the Census, *Historical Statistics*, p. 56.

22. From U.S., Congress, Senate, Immigration Commission, *Immigrants in Industries*, 2 vols., pt. 1: *Bituminous Coal Mining*, S. Doc. 633, 61st Cong., 2d sess., 1910, ser. 5667, 1:44 (hereafter referred to as Immigration Commission, *Report*).

23. UMWA *Proceedings*, 1916, pp. 556–57.

24. Aug. 25, 1904, p. 4; UMWJ, July 17, 1902, p. 4.

25. See *W. Va. Inspectors' Report, 1908*, p. 213.

26. Immigration Commission, *Report*, Table 94, p. 137.

27. Ibid., Table 128, p. 206.

28. Ibid., p. 232.

29. Compiled from *W. Va. Inspectors' Report, 1908*, p. 92, and Immigration Commission, *Report*, Table 148, p. 234.

30. Compiled from *Pa. Mines Report, 1912*, Table H, p. 76, and Immigration Commission, *Report*, Table 134, pp. 219–20. The data present two problems.

First, the miner sample on which the Pennsylvania data is based represents only about 25 percent of the total number of bituminous miners in that state. Second, some guesses had to be made in combining data from two sources. In the last column, for example, the 18.51 figure for Slovaks and Slovenians actually represents the general category "Slavs."

31. On the racism of the Immigration Commission, see Oscar Handlin, *Race and Nationality in American Life* (New York, 1957), chapt. 5, and I. A. Hourwich, *Immigration and Labor*, 2d ed. (New York, 1922).

32. Manning to 80 Senators and Congressmen (Feb. 13, 1919), RG 70, box 408, 031, 1919.

33. NSC *Proceedings*, Mining Section, 5 (1916):1010.

34. W. Jett Lauck, "The Bituminous Coal Miner and Coke Worker in Western Pennsylvania," *Survey* 26 (April 1, 1911):36; Korson, *Coal Dust*, p. 13. On miner education, see AMC *Proceedings* 19 (1916):562; Beckner, *Labor Legislation*, p. 301; UMWA *Proceedings*, 1916, p. 557; U.S., Congress, House, Committee on Mines and Mining, *Appropriations for Mining Schools*, Hearings before the Committee on Mines and Mining, House of Representatives, on H.R. 6063, 63d Cong., 2d sess., 1913, p. 4; and Van H. Manning, "Mine Safety Devices Developed by the United States Bureau of Mines," *Annual Report of the Board of Regents of the Smithsonian Institution, for the year ending June 30, 1916* (Washington, D.C., 1917), 71:535.

35. *Coal* 7 (Sept. 19, 1908):18. Also *Colliery Engineer* 34 (June 1914):698–99; *Coal and Coke Operator* 24 (March 1915):541; *Mines and Minerals* 28 (June 1908):516.

36. T. D. Lewis of the Lehigh Coal and Navigation Company, Lansford, Pa., in the discussion following Albert H. Fay, "Mine Accidents: English Speaking vs. Non-English Speaking Employees," NSC *Proceedings* 8 (1919):824–25. Also ibid., p. 116; 1 (1912):13; and J. H. Dague and S. J. Phillips, *Mine Accidents and Their Prevention* (New York, 1912). The broad outlines of the Americanization movement can be traced in John Higham, *Strangers in the Land*, Atheneum ed. (New York, 1971), pp. 234–63; and Edward George Hartmann, *The Movement to Americanize the Immigrant* (New York, 1967).

37. Holmes to Mitchell (Dec. 5, 1910), Mitchell papers, box 15, file 63; Manning to Senators and Congressmen (Feb. 13, 1919), RG 70, box 408, 031, 1919.

38. Cummins to Manning (Feb. 17, 1919), box 408, 031, 1919. That box contains much information on the Manning legislation.

39. Hartmann, *Movement to Americanize*, p. 263. The tension between education and social control is presented in Joel H. Spring, *Education and the Rise of the Corporate State* (Boston, 1972).

40. AMC *Proceedings* 16 (1913):74; MIIA *Proceedings*, 1914, pp. 142, 148; *Coal Age* 9 (Jan. 22, 1916):178–79, 180; (June 17, 1916):1048; *Coal Trade Bulletin* 17 (June 1, 1907):24. On dismissal, see the multipage document dated 1917, Employment Office, Muscoda Works, signed by Employment Officer J. L. McClaran. It is essentially a list of those dismissed by the company for any reason (Bittner papers, box 1, file "Correspondence, 1916"). Also Ben M. Selekman and Mary Van Kleeck, *Employes' Representation in Coal Mines* (New York, 1924), p. 111; *Coal Age* 9 (April 22, 1916): 699; MIIA *Proceedings*, 1908, p. 41.

41. "The Conservation of Coal in the United States," TAIME 40 (1909): 603; UMWJ, April 28, 1904, p. 4; Brophy, *Miner's Life*, pp. 39, 41; *Illinois Coal Report, 1908*, p. 4; *Ohio Inspectors' Report, 1912*, pp. 268, 300–301, 321, 369, 386; *Pa. Mines Report, 1904*, p. xi; and Ohio, *Report of Coal Mining Commission* (1914), pp. 330–31.

42. David Ross, *History of Coal Mining in Illinois*, An Address Delivered before the State Mining Institute in Annual Session at DuQuoin, Illinois, May 26th, 1916 (Springfield, n.d.), p. 10; *Coal Trade Bulletin* 22 (Jan. 15, 1910):44; and *Coal Age* 9 (June 17, 1916):1048.

43. Beckner, *Labor Legislation*, pp. 323–27; U.S., Department of Commerce and Labor, Bureau of Labor, "Labor Legislation in the United States" (Washington, D.C., 1904), 9:1428–30, esp. table facing p. 1428; UMWA *Proceedings*, 1905, p. 267; and Morris, *Plight*, pp. 148–49.

44. UMWJ, Jan. 28, 1904, p. 4; Jan. 26, 1911, p. 4; and Oct. 11, 1906, p. 4.

45. UMWA *Proceedings*, 1905, p. 29; 1906, pp. 45–48; 1908, pp. 307–8; Van Bittner to UMWJ (Dec. 12, 1907), Bittner papers, box 10, file "1910–1918"; 1908 House Bureau Hearings, pp. 36–37.

46. IWW, *Coal-Mine Workers*, p. 47; and Holmes to Mitchell (March 18, 1914), Mitchell papers, box 15, file 63.

47. *Mines and Minerals* 33 (Feb. 1913):348; 29 (Aug. 1908):18–19; *Black Diamond* 28 (Feb. 15, 1902):232; *Coal and Timber* 1 (March 1905):17; *Coal Trade Bulletin* 15 (July 2, 1906):44; and *Coal and Coke* 15 (July 1, 1908):9; MIIA *Proceedings*, 1915, pp. 35, 36, 38, 52–54; 1910, pp. 152–53; and 1919, pp. 18–19.

48. *Coal* 5 (March 14, 1907):18; 7 (July 9, 1908):18; *Coal and Coke Operator* 7 (Dec. 21, 1911):408; *Coal and Coke* 15 (July 1, 1908):9; and Beckner, *Labor Legislation*, p. 330.

49. Collins to Felts (Feb. 6, 1908), Collins papers, series 2; discussion of Fay, "Mine Accidents," p. 822; Brophy, *Miner's Life*, p. 117; Beckner, *Labor Legislation*, pp. 328–32; UMWJ, March 16, 1905, p. 1; Oct. 12, 1905, p. 4; Jan. 10, 1907, p. 4; Oct. 5, 1905, p. 1; April 25, 1907, p. 4; June 18, 1908, p. 2; U.S., Department of Labor, Bureau of Labor Statistics, *Labor Laws of the United States*, with Decisions of Courts Relating Thereto, *Bulletin* 148, pt. 2:1638, 1269; pt. 1: 699–702, 625–27; *Coal and Timber* 1 (March 1905):17; *Coal Trade Bulletin* 16 (Dec. 1, 1906):29–30; *Coal and Coke* 15 (July 1, 1908):9.

50. UMWJ, Sept. 9, 1909, p. 4; UMWA *Proceedings*, 1910, p. 571; 1909, pp. 434, 443; G. O. Virtue, "The Anthracite Mine Laborers," *Bulletin 13 of the U.S. Department of Labor* (Washington, D.C., 1897), pp. 768–69; UMWA, *District 12 Proceedings, 1911*, pp. 257–66; *Coal Trade Bulletin* 15 (July 2, 1906):44.

51. Copy of "Planks Proposed by Labor to the Republican Convention, June 18, 1908," in American Federation of Labor papers, State Historical Society, Wisconsin, series 11, box 7, file June 12–July 1; letter Samuel Gompers to F. M. King (April 3, 1908), AFL papers, series 11, box 7, file March 14–May 4; UMWJ, Jan. 22, 1914, p. 3; UMWA, International Executive Board, *Minutes*, meeting of Feb. 4, 1912, p. 176 (in MS). For UMWA interest in safety outside the coal mines, see UMWJ, March 1, 1917, p. 12; June 14, 1917, p. 9; and Sept. 13, 1917, p. 12.

52. Edward A. Wieck, *The American Miners' Association* (New York, 1940), pp. 203–4; Arthur E. Suffern, *The Coal Miners' Struggle for Industrial Status* (New York, 1926), pp. 21–22, 26, 30–31, 36–41; Chris Evans, *History of the*

UMWA, 1:37–38, 98, 103, 105, 199, 224, 262–63, 403, 492; and D. R. Jones, *The Mining Conflict* (Pittsburgh, 1880). K. Austin Kerr is more convinced than I am of the high priority of safety in the affairs of these early unions. See his "State Regulation of Coal Mines in the Nineteenth Century" (paper delivered at the Business History Conference, Northwestern University, Feb. 28, 1975).

53. Baratz, *Union and Coal Industry*, p. 52; Evans, *History of the UMWA*, 2:75, 78–79, 85–86, 119, 154, 189, 267, 706. The membership figures are from Mitchell papers, box 124, no file number.

54. [*sic*] UMWA, *Minutes of the 19th Annual Convention of District Number 2, Du Bois, Pennsylvania, March 24–30, 1908* (n.p., n.d.), pp. 10, 12.

55. UMWA, International Executive Board, *Minutes*, meeting of May 24, 1910, pp. 370, 366 (in MS); and *Minutes of the International Executive Board*, March 30, 1908, meeting, Indianapolis (n.p., n.d.).

56. UMWA *Proceedings*, 1910, pp. 89, 114; and UMWJ, Jan. 27, 1910, p. 4.

57. UMWA *Proceedings*, 1912, pp. 20, 27; Mitchell papers, box 124, no file number; and Selig Perlman and Philip Taft, *Labor Movements*, vol. 4: *History of Labor in the United States, 1896–1932*, ed. John R. Commons (New York, 1935), p. 30.

58. UMWA *Proceedings*, 1916, p. 963 (Jones); 1912, p. 466; 1916, pp. 23, 204 (White); and 1914, pp. 50–51, 57.

59. UMWJ, Aug. 5, 1915, p. 4; UMWA *Proceedings*, 1918, pp. 47–50.

60. Marion Dutton Savage, *Industrial Unionism in America* (New York, 1922), pp. 106–7; UMWJ, Dec. 15, 1918, p. 2.

61. (Dec. 30, 1938), AALL papers, Record Group A–1, box 69.

62. Hanraty to Mitchell (June 25, 1909), Mitchell papers, box 23, file 111. The 1912, 1916, and 1924 Interstate Contracts are printed in full in the Appendix to Louis Bloch, *Labor Agreements;* Interstate Joint Conference, *Proceedings*, 1912, pp. 15–16.

63. Joint Convention of the Illinois Coal Operators' Association and the UMWA (District 12), held at Peoria, Illinois, February 24 to March 13, 1902, *Proceedings* (n.p., n.d.), pp. 74, 77 (hereafter referred to as Joint Convention, District 12, *Proceedings*); ibid., *1903*, p. 198; *Coal Trade Bulletin* 21 (Nov. 1, 1909):33.

64. UMWJ, June 7, 1906, p. 5.

65. 1908 Kanawha Agreement, quoted in *W. Va. Hearings, 1909*, p. 630.

66. Joint Convention, District 12, *Proceedings, 1902*, pp. 92, 17; *1903*, p. 113.

67. Ibid., *1904*, pp. 189–90, also pp. 176–79, 188; *1903*, p. 204; *1906*, p. 99; *Coal Trade Bulletin* 22 (March 15, 1910):48. This discussion of contract provisions is compiled from the following sources: Kanawha Agreement, 1908, *W. Va. Hearings, 1909*, pp. 629–31; Southeastern Kentucky and East Tennessee Agreement, 1920, Bittner papers, box 1, file "Correspondence, April–July, 1920"; Michigan Scale Agreement, 1906, UMWJ, July 12, 1906, p. 2; Montana Agreement, 1907, UMWJ, Oct. 3, 1907, p. 2; Indiana Agreement, 1906, UMWJ, June 14, 1906, p. 5; Joint Interstate Agreement of Pittsburg, Kansas, 1903, Mitchell papers, box 119, no file number; Hocking Valley Agreement, 1908, UMWJ, May 21, 1908, p. 8; Hocking Valley Agreement, 1906, UMWJ, Aug. 9, 1906, p. 2; District 21 Agreement, 1903, UMWJ, Sept. 10, 1903, p. 8; Central Kentucky Agreement, 1904, UMWJ, July 7, 1904, p. 6; Illinois (District 12) Agreement, 1903, Mitchell papers, box 119, no file number; Illinois Agreement, 1904, U.S., Department of Commerce and Labor, Bureau of Labor, *Bulletin* 52 (Washington, D.C., 1904),

pp. 642–46, and UMWJ, April 14, 1904, p. 5; Illinois Agreement, 1906, UMWJ, June 7, 1906, p. 5; Illinois Agreement, 1912, in Bloch, *Labor Agreements*, pp. 368, 372, 397, 371, 407 (all page numbers refer to appendix), and for Kansas agreement, Chairman, Southwestern Coal Operators' Association to Manning (June 8, 1917), RG 70, box 330, 711, 1917.

68. Joint Convention, District 12, *Proceedings, 1903*, p. 95.

69. 1920 Southeastern Kentucky and East Tennessee Agreement, in Bittner papers; Joint Convention, District 12, *Proceedings, 1904*, p. 171; Bureau of Labor *Bulletin* 52, p. 645.

70. UMWJ, May 13, 1909, p. 1; and Goodrich, *Miner's Freedom*, pp. 86–87. In 1941 Edward Wieck (*Preventing Fatal Explosions in Coal Mines* [New York, 1942], p. 107) claimed that agreements did not include safety provisions and concluded that questions of safety were not, therefore, handled by the pit committees. Relying on Wieck for his information, historian Ludwig Teleky (*Factory and Mine Hygiene*, pp. 81–82) concludes that the agreement "hindered rather than promoted the progress of safety measures." Neither claim seems justified. Illinois miners and operators worked out a different kind of adjudication mechanism, contractually endowing a tripartite commission (the Illinois Mining Investigation Commission) with the rights and duties of preparing all new mining-safety legislation. UMWA, *District 12 Proceedings, Reconvened Session, 1910*, pp. 7–12.

71. Suffern, *Coal Miners' Struggle*, pp. 255 (chart), 257–58, 264 (chart), 229, 242–45, 377–78; Bloch, *Labor Agreements*, pp. 125–26, 156–57. For one example of a contract dispute over a safety question—this one involving the use of machinery—see President, District 5, UMWA to William Green (Aug. 3, 1912) and W. H. Howarth to Bittner (June 1, 1912), both in Bittner papers, box 1, file "Correspondence 1912"; records of disputes under contract can also be found in the *Monthly Bulletin* of the Illinois Coal Operators' Association.

72. Industrial Commission, *Report*, 12:91; Earl A. Saliers, *The Coal Miner* (n.p., 1912), p. 12; Samuel Gompers, "Strikes and the Coal-Miners," *Forum* 24 (Sept. 1897):27–33; Winthrop D. Lane, *Civil War in West Virginia* (New York, 1921); and Victor Hicken, "The Virden and Pana Mine Wars of 1898," *Journal of the Illinois State Historical Society* 52 (Summer 1959):263–78.

73. U.S., Congress, Senate, Anthracite Coal Strike Commission, *Report to the President on the Anthracite Coal Strike of May–October, 1902, by the Anthracite Coal Strike Commission*, S. Doc. 6, 58th Cong., special sess., 1903, ser. 4556, 1:93.

74. U.S., Congress, House, *Report on the Miners' Strike in Bituminous Coal Field in Westmoreland County, Pa., in 1910–11*, H. Doc. 847, 62d Cong., 2d sess., 1912, ser. 6279, pp. 134–36; U.S., Congress, Senate, *A Report on Labor Disturbances in the State of Colorado, from 1880 to 1904, Inclusive*, S. Doc. 122, 58th Cong., 3d sess., 1905, ser. 4765, 3:330–31; and S. Doc. 415, *Industrial Relations, on the Colorado Coal Strike, 1913–14*, p. 7047.

75. UMWJ, July 16, 1908, p. 8; Dec. 3, 1908, p. 5; UMWA *Proceedings*, 1909, pp. 62–63; UMWA, International Executive Board, *Minutes*, meetings of Jan. 22 and June 23, 1908, pp. 91, 135; Illinois Coal Operators' Association, *Monthly Bulletin* 3 (Feb. 1911):47; 1 (Nov. 1909):250; 1 (July 1909):163; *Coal Age* 2 (Oct. 12, 1912):501.

76. Mitchell, *Organized Labor* (Philadelphia, 1903), pp. 150, 9.

77. Interstate Joint Conference, *Proceedings*, 1908, p. 83; UMWA, International Executive Board, *Minutes*, meeting of May 21, 1910, p. 193, and May 20, 1910, pp. 164–65 (in MS).

78. Woodrow Wilson papers, series 4, box 384, file 2097; J. T. Ryan, Vice President, Mine Safety Appliances Co., to Manning (July 28, 1916); Ryan to Manning (Oct. 13, 1916), RG 70, box 188, 021.33, 1916; H. M. Wilson to Manning (Aug. 10, 1916), box 238, 711, 1916; and Manning to W. D. Ryan (March 14, 1917), box 332, 821, 1917.

79. U.S., Congress, House, *Permissible Explosives Tested prior to January 1, 1911, and Precautions to Be Taken in Their Use*, by Clarence Hall, Bureau of Mines, H. Doc. 1405, 61st Cong., 2d sess., 1911, ser. 6069; MIIA *Proceedings*, 1910, p. 68 (Hall quote); Bureau of Mines, *Third Annual Report* (1913), p. 16; Fay, "Mine Accidents and Uniform Records," p. 495; and Andrews, "Needless Hazards," p. 27.

80. UMWA *Proceedings*, 1910, pp. 555–56; 1914, p. 1128; UMWJ, Oct. 21, 1909, p. 1; UMWA, International Executive Board, *Minutes*, meetings of March 31 and June 25, 1909; Mitchell to Reverend Friend (March 25, 1911), Mitchell papers, box 32, file 167.

81. UMWA, International Executive Board, *Minutes*, meeting of Sept. 24, 1909; also UMWJ, June 17, 1909, p. 1.

82. UMWJ, Oct. 7, 1909, p. 4.

83. 21 (Sept. 21, 1909):26; and *Coal and Coke Operator* 9 (Sept. 16, 1909): 168.

84. UMWA, *District 12 Proceedings, 1911*, p. 34.

85. UMWA *Proceedings*, 1916, p. 31; 1902, p. 45; *W. Va. Hearings, 1909*, pp. 629–30; TAIME 41 (1910):478; AMC *Proceedings* 14 (1911):227; S. Sanford, unpublished summary of fieldwork done for the Geological Survey entitled "Wastes in Bituminous Coal Mining," and John Shober Burrows, "Report of the Investigation during 1907 of Waste in Bituminous Coal Mining," also unpublished, in RG 70, box 38, 616, 1910–11; and UMWJ, May 13, 1915, pp. 12–13.

86. Interstate Joint Conference, *Proceedings*, 1906, p. 41, also pp. 85, 98.

87. AMC *Proceedings* 14 (1911):65–66, 92, 234–35; UMWJ, Feb. 11, 1909, p. 3; *Bulletin* 1, Juanita Coal and Coke Company, Bowie, Colo., Sept. 1, 1907, typewritten copy, RG 70, box 35, 474, 1910–11.

88. Mitchell to Holmes (July 2, 1908), Mitchell papers, box 15, file 63; UMWJ, March 2, 1905, p. 4; UMWA *Proceedings*, 1906, pp. 122, 139; 1904, p. 140.

89. UMWA *Proceedings*, 1905, p. 172; also 1909, pp. 62–63; UMWJ, March 31, 1910, p. 4; Interstate Joint Conference, *Proceedings*, 1912, pp. 56, 70, 133, 114, 119, 140.

90. *Fuel* (May 8, 1905), p. 18; (May 15, 1905), p. 58. The Illinois shot-firing controversy may be traced in Beckner, *Labor Legislation*, pp. 344 ff., various numbers of the UMWJ, and in the Mitchell papers, box 101, no file number. On the subject of the dangers of being a shot-firer and the miner reaction to this danger, see UMWJ, March 2, 1905, p. 4; Oct. 29, 1903, p. 1; Jan. 21, 1904, p. 3.

91. MIIA *Proceedings*, 1914, p. 163.

92. UMWJ, Jan. 21, 1909, p. 12.

93. The statistics were compiled in 1908 by John Mitchell, but the results appear valid if other statistics are used. Compare U.S., Coal Commission, *Report*, transmitted pursuant to the act approved September 22, 1922, 5 parts, 1925, S.

Doc. 195, 68th Cong., 2d sess., ser. 8402–3, pt. 3: "Bituminous Coal, Labor and Engineering Studies," p. 1658. See Mitchell, "Conservation in the Coal Industry," in AMC *Proceedings* 11 (1908):187–89; and typewritten manuscript in Mitchell papers, box 19, file 87. Admittedly, the statistics do not prove that the union was the cause of the safer mines. It is entirely possible that organization is not an independent variable but is instead related to some other factor of crucial significance such as labor force composition, company size, general political and social climate, or the state of development of the coalfield.

94. UMWJ, Nov. 26, 1914, p. 4. Also UMWJ, Aug. 19, 1915, p. 4; *Ohio Inspectors' Report, 1907*, p. 268; UMWA, *District 12 Proceedings, 1910*, p. 27.

CHAPTER 5

1. Korson, *Coal Dust*, p. 255.

2. To Eugene Debs (Dec. 8, 1909), in Germer papers, box 1, Correspondence 1901–32, file "correspondence 1901–1910."

3. Korson, *Coal Dust*, p. 250.

4. [Mary Harris] Jones, *Autobiography*, p. 200.

5. *Pa. Mines Report, 1909*, p. lx.

6. *Pa. Mines Report, 1904*, p. 289; *W. Va. Inspectors' Report, 1906*, p. 256.

7. Graebner, "Great Expectations." Although demand for coal underwent major seasonal fluctuations, supply was capable only of easy expansion. Shutting down a coal mine was so expensive that it was often cheaper to continue production at a loss. Each period of prosperity in the coal business was followed by great increases in capacity.

8. *Coal Trade Bulletin* 18 (March 2, 1908):26, and 25 (June 15, 1911):21; *Coal* 7 (Jan. 23, 1908):19; *Coal and Coke Operator* 10 (Jan. 20, 1910):41.

9. (Dec. 24, 1909), Fleming papers, Aug.–Dec. 1909 box.

10. (Nov. 27, 1908), Fleming papers, Sept. 1908–Jan. 1909 box.

11. (Nov. 22, 1909), Fleming papers, Aug.–Dec. 1909 box.

12. Collins to T. L. Felts (Feb. 6, 1908), Justus Collins papers, West Virginia University Library, Morgantown, series II. Collins opened his first coal mine in the Pocahontas-Flat Top coalfield of southern West Virginia, and later operated mines in New River, Tug River, and the Winding Gulf coalfields. He headed a coal sales agency and a cement company, speculated in coal and timber lands, and played a central role in organizing the Tug River Coal Operators' Association, the Winding Gulf Operators' Association, and the Smokeless Coal Operators' Association.

13. (Jan. 15, 1909), Collins papers, series I, box 1, folder 21.

14. Collins to Mann (May 25, 1916, and June 1, 1917), Collins papers, series I, box 15, folder 103.

15. To A. M. Herndon (Nov. 20, 1911), Winding Gulf Colliery Company papers, West Virginia University Library, Morgantown, box 1.

16. Collins to Jairus Collins (Jan. 13, 1909), Collins papers, series I, box 1, folder 21; George Wolfe to Collins (Aug. 22, 1916), Collins papers, series I, box 14, folder 97; Collins to Herndon (May 15, 1911), Winding Gulf papers, box 1, folder 5; Wolfe to Collins (March 5, 1915), Winding Gulf papers, box 2,

folder 1; and Wolfe to Collins (Aug. 11 and May 10, 1915), Winding Gulf papers, box 2.

17. Selekman and Van Kleeck, *Employes' Representation*, pp. 386–87.

18. Ibid., pp. 22–23, 63–64, 76, 137; and Suffern, *Coal Miners' Struggle*, pp. 286–87.

19. Selekman and Van Kleeck, *Employes' Representation*, pp. 184, 145, 188–89. Also Irving Bernstein, *The Lean Years* (Baltimore, 1960), pp. 162–64.

20. William Clifford to Fleming (Dec. 24, 1907), Fleming papers, box 36.

21. William H. Tolman and Leonard B. Kendall, *Safety* (New York, 1913), p. 172, also pp. 171, 173–77; Goodrich, *Miner's Freedom*, p. 126; *Mines and Minerals* 30 (Oct. 1909):174; John A. Garraty, "The United States Steel Corporation versus Labor: The Early Years," *Labor History* 1 (Winter 1960):20–21, 35.

22. Bennett and Graebner, "Safety First."

23. UMWJ, Aug. 1, 1901, p. 4; NSC *Proceedings* 7 (1918):1025; Thomas Lynch to Holmes (April 30, 1909), RG 70, box 35, 474, 1910–11; and H. C. Frick Coke Company, *Safety the First Consideration* (n.p., 1916).

24. NSC *Proceedings*, Mining Section, 1 (1912):83.

25. *Coal Age* 5 (Jan. 24, 1914):169–70; *Coal Trade Bulletin* 21 (Oct. 15, 1909):30; *Mines and Minerals* 30 (Oct. 1909):174; *Coal and Coke Operator* 13 (Oct. 5, 1911):216.

26. U.S., Congress, Senate, Commission on Industrial Relations, *Industrial Relations*, Final Report and Testimony, 11 vols., S. Doc. 415, 64th Cong., 1st sess., 1916, ser. 6935, 7:6463; *Coal Trade Bulletin* 24 (May 15, 1911):33; 1908 House Bureau Hearings, pp. 32, 33; NSC *Proceedings* 1 (1912):83; AMC *Proceedings* 18 (1915):74; and E. A. Holbrook to Alice Willeford (April 30, 1919), RG 70, box 408, 031, 1919.

27. NSC *Proceedings* 5 (1916):104.

28. Ibid., 8 (1919):20.

29. Ibid., 1 (1912):243.

30. Coal Mining Institute of America (CMIA), *Proceedings, 1904*, p. 158; S. Rept. 692, *Establishing Bureau of Mines in Interior Department, 1908*, p. 21; *Coal* 7 (Oct. 8, 1908):18; *Colliery Engineer* 35 (Nov. 1914):209.

31. Roy Lubove, "Workmen's Compensation and the Prerogatives of Voluntarism," *Labor History* 8 (Fall 1967):261; James H. Boyd, *Workmen's Compensation*, 2 vols. (Indianapolis, 1913), 1:54; William H. Shockley, "The Economic and Social Influence of Mining with Special Reference to the United States," *Transactions of the International Engineering Congress* (1915), pp. 1–66. The employer could argue under the fellow servant rule that he was not responsible financially for any accident caused by a fellow employee of the injured; under the assumption of risk rule, that the employee assumed all risks involved in the employment provided the employer took reasonable care; and under contributory negligence, that he was not financially responsible for any accident for which the employee was at all responsible.

32. E. H. Downey, *History of Work Accident Indemnity in Iowa* (Iowa City, 1912), p. 86.

33. Mitchell, *The Wage Earner and His Problems* (Washington, D.C., 1913), p. 53.

34. Downey, *Work Accident Indemnity*, pp. 112, 119, 107; Beckner, *Labor*

Legislation, pp. 438 ff.; Peter Roberts, *Anthracite Coal Communities* (New York, 1904), pp. 267–69; and Harry A. Millis and Royal E. Montgomery, *Labor's Risks and Social Insurance* (New York, 1938), pp. 190–205.

35. James Weinstein, "Big Business and the Origins of Workmen's Compensation," *Labor History* 8 (Spring 1967):166, 168–69, 170 n, 174; UMWA *Proceedings*, 1910, p. 575; 1912, p. 419; AMC *Proceedings* 11 (1908):62, 108; 13 (1910): 174; 14 (1911):92, 112–18; 15 (1912):109, 139; and 18 (1915):163.

36. Lubove, "Workmen's Compensation," pp. 259, 264, 261; and Weinstein, "Big Business," p. 164.

37. Mitchell, *Wage Earner*, p. 48; Goodrich, *Miner's Freedom*, p. 111.

38. TAIME, 1917, p. 553; W. G. Cowles to Holmes (Feb. 4, 1915); Holmes to Wilson (Feb. 24, 1915); Rice to Cowles (March 27, 1915); and Wilson to Manning (Feb. 15, 1915), all in RG 70, box 169, 455, 1915.

39. TAIME, 1917, p. 551, 553; AMC *Proceedings* 18 (1915):160; *Transactions of the International Mining Congress, 1915*, p. 172. For a detailed explanation of the determination of rates, see Wilson, "Inspection and Schedule Rating for Coal Mine Insurance," reprinted from *Proceedings of Casualty Actuarial and Statistical Society of America* 2, pt. 1:39–48, copy in RG 70, box 238, 711, 1916.

40. Rice to Manning memorandum (Feb. 9, 1915) and Rice to Cowles (March 27, 1915), both in RG 70, box 169, 455, 1915.

41. To Wilson (March 14, 1916), RG 70, box 238, 711, 1916.

42. Wilson to Rice (May 19, 1916) and Rice to Wilson (May 29, 1916), RG 70, box 238, 711, 1916.

43. D. J. Parker to Acting Chief Mining Engineer (April 8, 1919), and Holbrook to Director (April 15, 1919), RG 70, box 442, 454, 1919; Wilson to Manning (Aug. 10, 1916), RG 70, box 238, 711, 1916; Wilson to Manning (Sept. 12, 1916) and D. J. Parker to Manning (Nov. 9, 1917), RG 70, box 238, 711, 1916. According to one estimate, a company complying with rescue and first-aid requirements would save $840 on an annual payroll of $200,000. See also file 031, in box 408, 1919.

44. Wilson to Manning (Sept. 12, 1916), Acting Director to Wilson (Aug. 21, 1916), Rice to Manning (Jan. 15, 1917), RG 70, box 238, 711, 1916; unsigned to D. J. Parker (April 1, 1919), RG 70, box 442, 454, 1919.

45. Wilson to T. H. Devlin (Nov. 6, 1917), RG 70, box 330, 711, 1917; Wilson to Manning (Sept. 12, 1916), RG 70, box 238, 711, 1916; Wilson to Bureau Director (May 17, 1919), RG 70, box 403, 002, 1919; *Coal Age* 9 (May 27, 1916): 937; Wilson to Manning (Sept. 20, 1916) and J. J. Rutledge to Manning (Sept. 22, 1916), RG 70, box 238, 711, 1916; Wilson to Manning (Feb. 8, 1917) and Wilson to Governor Arthur Capper of Kansas (Aug. 15, 1917), RG 70, box 330, 711, 1917.

46. AMC *Proceedings* 14 (1911):264; Graebner, "Great Expectations."

47. Holmes to Fleming (Dec. 24, 1909), Fleming papers, Aug.–Dec. 1909 box.

48. Holmes to Mitchell (undated, probably Nov. 1909), Mitchell papers, box 15, file 63.

49. Copy in Collins papers, series III, box 37, folder 1; Suffern, *Coal Miners' Struggle*, pp. 168–69, 173; Bloch, *Labor Agreements*, pp. 60, 67 ff.; AMC *Proceedings* 19 (1916):77, 186; 18 (1915):99–100.

50. Beckner, *Labor Legislation*, pp. 294 ff.; W. R. J. Zimmerman, Secretary

of the Smokeless Coal Operators' Association, to Collins (June 30, 1913), Collins papers, series III, box 37, folder 1; CMIA *Proceedings; Coal Trade Bulletin* 11 (Oct. 1, 1904):46; *Coal* 6 (July 25, 1907):17; *Coal and Coke Operator* 11 (Dec. 15, 1910):830.

51. *Black Diamond* 48 (Jan. 27, 1912):35.

52. Fleming to Neil Robinson (Dec. 31, 1909), Fleming papers, box Aug.–Dec. 1909; *Coal and Coke Operator* 13 (Dec. 14, 1911):388, 381; Suffern, *Coal Miners' Struggle*, pp. 168–70; Suffern, *Conciliation and Arbitration*, pp. 128–34.

53. H. Foster Bain to Holmes (Dec. 9, 1912), RG 70, box 41, 003, 1912; NSC *Proceedings* 5 (1916):7, 47–49; 6 (1917):1332, 169; 7 (1918):187–88; Palmer to Holmes (July 19, 1912), RG 70, box 65, 445.50, 1912; and T. T. Read to B. F. Tillson (Nov. 6, 1919), RG 70, box 424, 089.20, 1919.

54. Mitchell, *Organized Labor*, p. 149.

CHAPTER 6

1. P. 135.

2. Ibid., pp. 134–35.

3. Ibid., pp. 196–97.

4. "Social Tensions and the Origins of Progressivism," *Journal of American History* 56 (Sept. 1969):336–39.

5. *The Education of Henry Adams: An Autobiography* (Boston, 1918).

6. Carl Vrooman, "The Ultimate Issue Involved in Railroad Accidents," *Arena* 39 (1908):19; Lucian W. Chaney and Hugh S. Hanna, *The Safety Movement in the Iron and Steel Industry, 1907–1917*, U.S., Department of Labor, Bureau of Labor Statistics, *Bulletin* 234 (Washington, D.C., 1918), p. 16.

7. Don D. Lescohier, *Working Conditions*, Vol. 3: *History of Labor in the United States, 1896–1932*, ed. John R. Commons (New York, 1935), p. 367; Shockley, "Economic and Social Influence of Mining," p. 30; Lew R. Palmer, "History of the Safety Movement," *Annals of the American Academy of Political and Social Science* 123 (Jan. 1926):9–12.

8. Bureau of Mines, *Sixth Annual Report* (1916), p. 68.

9. William H. Tolman and Leonard B. Kendall, *Safety* (New York, 1913), p. 396, and Palmer, "Safety Movement," p. 13; AMC *Proceedings* 19 (1916):41–45; 22 (1919):293; U.S., Department of the Interior, Bureau of Mines, *The Joseph A. Holmes Safety Association and Its Awards*, by D. Harrington, Louise Pedlow, and Anna P. Brown, *Bulletin* 421 (Washington, D.C., 1940), pp. 3–4; RG 70, box 441, 445, 1919, and box 189, 022.49, 1916; Mitchell papers, box 4, file 158; RG 70, box 8, 1910–11, program for 1911 AALL meeting.

10. Mitchell papers, box 10, file 44, and box 76, file 45, clipping from *Fuel* magazine.

11. "American Academy on Work-Accidents," *Survey* 26 (April 15, 1911):107–8; Joseph A. Holmes, "Government Measures to Increase Mine Safety," *Annals of the American Academy of Political and Social Science* 38 (1911):112–14, and the whole volume. See also Bennett and Graebner, "Safety First."

12. RG 70, box 151, 022.74, 1915; George Bradshaw, *Safety First* (New York, 1913); Lucian W. Chaney, "Safety as Promoted by Federal Bureaus," NSC *Proceedings* 2 (1913); Dupree, *Science*, pp. 269–70.

13. Adolph O. Eberhardt, "Conservation," AMC *Proceedings* 12 (1909):162–67.

14. Charles Richard Van Hise, *The Conservation of Natural Resources* (New York, 1910), p. 364.

15. NSC *Proceedings* 4 (1915):666–67.

16. Samuel P. Hays, *Conservation*, pp. 1–4; J. Leonard Bates, "Fulfilling American Democracy: The Conservation Movement, 1907 to 1921," *Mississippi Valley Historical Review* 44 (June 1957):29–57. The ends/means dichotomy is also drawn in Ronald G. Walters, "The Erotic South: Civilization and Sexuality in American Abolitionism," *American Quarterly* 25 (May 1973):177–201.

17. D. R. Kennedy, "Efficiency in Safety Work," NSC *Proceedings* 3 (1914): 136; Haber, *Efficiency and Uplift* (Chicago, 1964), pp. ix–x.

18. See Donald L. Kemmerer, "The Changing Pattern of American Economic Development," *Journal of Economic History* 16 (Dec. 1956):575–89. The appearance of scientific management is often taken as evidence of a labor shortage. It is becoming increasingly evident, however, that labor force control was scientific management's basic objective. See Katherine Stone, "The Origins of Job Structures in the Steel Industry," *Review of Radical Political Economics* 6 (Summer 1974):113–73.

19. See Donald K. Pickens, *Eugenics and the Progressives* (Nashville, Tenn., 1968), pp. 66–67; John S. Haller, Jr., *Outcasts from Evolution: Scientific Attitudes of Racial Inferiority, 1859–1900* (Urbana, Ill., 1971); John Higham, *Strangers in the Land*, chapts. 6, 7, 9. Kate Holladay Claghorn, *The Immigrant's Day in Court* (New York, 1923), pp. 19–21, 102; and Joseph M. Hawes, *Children in Urban Society: Juvenile Delinquency in Nineteenth-Century America* (New York, 1971), p. 141.

20. "Duties of the Safety Engineer," NSC *Proceedings* 6 (1917):91; James Weinstein, *The Corporate Ideal in the Liberal State, 1900–1918* (Boston, 1968), pp. 19, 47; Haber, *Efficiency and Uplift*, pp. 18–19.

21. Joel H. Spring, *Education and the Corporate State;* James H. Timberlake, *Prohibition and the Progressive Movement, 1900–1920* (Cambridge, Mass., 1963).

22. Rothman, *The Discovery of the Asylum: Social Order and Disorder in the New Republic* (Boston, 1971); Katz, *Class, Bureaucracy, and Schools: The Illusion of Educational Change in America* (New York, 1971); Walters, "Erotic South."

23. David Musto, *The American Disease: Origins of Narcotic Control* (New Haven, 1973), p. 65; Spring, *Education and the Corporate State;* Lesy (New York, 1973), conclusion; and works by Higham, Pickens, and Haller cited above.

24. William Ryan, *Blaming the Victim* (New York, 1971), and Barbara Ehrenreich and Deirdre English, *Complaints and Disorders: The Sexual Politics of Sickness* (Old Westbury, N.Y., 1973), pp. 51–54.

25. 1908 House Bureau Hearings, p. 62; Wiebe, *The Search for Order, 1877–1920* (New York, 1967), pp. 151, 143–45, 193–95; and Edwin L. Earp, *The Social Engineer* (New York, 1911), p. 13.

26. UMWJ, Oct. 28, 1909, p. 4.

27. NSC *Proceedings* 7 (1918):84; and Van Schaack, "Duties of the Safety Engineer," p. 81.

28. Russel B. Nye, *Midwestern Progressive Politics* (East Lansing, Mich., 1951), p. 201.

29. Quoted in UMWJ, June 1, 1911, p. 1. See also A. M. Simons, *Wasting Human Life* (n.p., n.d.).

30. Louise C. Wade, *Graham Taylor: Pioneer for Social Justice, 1851–1938* (Chicago, 1964), p. 205.

31. Walter I. Trattner, *Crusade for the Children: A History of the National Child Labor Committee and Child Labor Reform in America* (Chicago, 1970), pp. 11–12.

32. John D. Buenker, "Urban Immigrant Lawmakers and Progressive Reform in Illinois," in *Essays in Illinois History: In Honor of Glenn Huron Seymour*, ed. Donald F. Tingley (Carbondale, 1968), p. 60.

33. J. Joseph Huthmacher, "Urban Liberalism and the Age of Reform," *Mississippi Valley Historical Review* 49 (Sept. 1962):231–41.

34. Oscar E. Anderson, Jr., *The Health of a Nation: Harvey W. Wiley and the Fight for Pure Food* (Chicago, 1958), and Trattner, *Crusade*, pp. 120 ff.

35. I am grateful to Gerald D. Nash's "Bureaucracy and Economic Reform: The Experience of California, 1899–1911," *Western Political Quarterly* 13 (Sept. 1960):678–91, for direction to the relevant literature, including O. Douglas Weeks, "Initiation of Legislation by Administrative Agencies," *Brooklyn Law Review* 9 (1939–1940):117–31; Elizabeth McK. Scott and Belle Zeller, "State Agencies and Lawmaking," *Public Administration Review* 2 (Summer 1942): 205–20; W. Brooke Graves, "Federal Leadership in State Legislation," *Temple Law Quarterly* 10 (July 1936):384–405; Edwin E. Witte, "Administrative Agencies and Statute Lawmaking," *Public Administration Review* 2 (Spring 1942): 116–25. See also Gerald D. Nash, *State Government and Economic Development: A History of Administrative Policies in California, 1849–1933* (Berkeley, 1964).

36. Weinstein, *Corporate Ideal*, p. ix.

37. *Age of Reform*, pp. 233, 216; Gabriel Kolko, *Railroads and Regulation, 1877–1916* (Princeton, 1965), p. 2.

38. Kolko, *Railroads*, p. 238; *Triumph of Conservatism* (Chicago, 1967), p. 6.

39. Keach Johnson, "Iowa Dairying at the Turn of the Century: The New Agriculture and Progressivism," *Agricultural History* 45 (April 1971):95–110; Weinstein, *Corporate Ideal*, chapt. 3; and Robert D. Cuff, *The War Industries Board: Business-Government Relations during World War I* (Baltimore, 1973).

40. "Business Disunity and the Progressive Movement, 1901–1914," *Mississippi Valley Historical Review* 44 (March 1958):665.

41. Weinstein, *Corporate Ideal*, chapt. 4; Hays, "The Politics of Reform in Municipal Government in the Progressive Era," *Pacific Northwest Quarterly* 55 (Oct. 1964):157–69.

42. Weinstein, *Corporate Ideal*, chapt. 2; Lubove, *The Struggle for Social Security, 1900–1935* (Cambridge, Mass., 1968), p. 49; and Wesser, "Conflict and Compromise: The Workmen's Compensation Movement in New York, 1890s–1913," *Labor History* 11 (Summer 1971):345–72.

43. *Businessmen and Reform: A Study of the Progressive Movement*, paperback ed. (Chicago, 1968), pp. 211–12.

44. *Health of a Nation* and *Prohibition and the Progressive Movement.*

45. The causes of business disunity discussed by Wiebe—function, geography, size—seem unrelated to differences within the coal industry, which were

largely a product of competitive markets. See Wiebe, "Business Disunity."

46. Definitions are in Kolko, *Triumph of Conservatism*, p. 3.

47. *Search for Order*, p. 12; also pp. 67, 74, 112.

48. Hays, *Conservation*, in the preface to the Atheneum edition.

49. Johnson, "Iowa Dairying"; Hays, "The Politics of Reform in Municipal Government"; Kolko, *Triumph of Conservatism*, pp. 98–108; Gerald E. Markowitz and David Karl Rosner, "Doctors in Crisis: A Study of the Use of Medical Education Reform to Establish Modern Professional Elitism in Medicine," *American Quarterly* 25 (March 1973):83–107; and Jerry Israel, ed., *Building the Organizational Society: Essays on Associational Activities in Modern America* (New York, 1972).

50. William Graebner, "The Coal-Mine Operator and Safety: A Study of Business Reform in the Progressive Period," *Labor History* 14 (Fall 1973):483–505; Kolko, *Triumph of Conservatism*, p. 6; Wiebe, *Search for Order*, pp. 198–220; Dupree, *Science*, p. 271.

51. Boulding, *The Organizational Revolution: A Study in the Ethics of Economic Organization* (New York, 1953); Wiebe, *Search for Order*, pp. 222–23; and Hays, introduction to Israel, *Building the Organizational Society*, p. 15.

52. William Graebner, review of Hace Sorel Tishler, *Self-Reliance and Social Security, 1870–1917* (Port Washington, 1971), in *Journal of Social History* 6 (Spring 1973):378, 380 n.

53. Trattner, *Crusade*, pp. 120, 136, 142; Jeremy P. Felt, *Hostages of Fortune: Child Labor Reform in New York State* (Syracuse, 1965), pp. 77, 195–99; and Sheldon Hackney, *Populism to Progressivism in Alabama* (Princeton, 1969), pp. 247–48.

54. Forrest A. Walker, "Compulsory Health Insurance: The Next Great Step in Social Legislation," *Journal of American History* 56 (Sept. 1969):290–304; Musto, *American Disease*, p. 65; and Timberlake, *Prohibition and the Progressive Movement*.

55. *The American Partnership: Intergovernmental Co-operation in the Nineteenth-Century United States* (Chicago, 1962), pp. 297–98, 323–25, 337.

56. W. Brooke Graves, *American Intergovernmental Relations: Their Origins, Historical Development, and Current Status* (New York, 1964), pp. 798, 785. A fuller account of the uniformity movement is presented in William Graebner, "Federalism and the Progressive Period: A Structural Interpretation of Reform" (paper delivered at the Sixty-eighth Annual Meeting of the Organization of American Historians, Boston, April 17, 1975).

57. Graves, *American Intergovernmental Relations*, p. 584; Graves, *Uniform State Action: A Possible Substitute for Centralization* (Chapel Hill, 1934), pp. 46–48, 53–56, 79, 293–96.

58. *Hostages of Fortune*, p. 224; and Trattner, *Crusade*, pp. 40–41, 47–48, 55–60, 88, 116, 122, 242–43. See also Jesse Thomas Carpenter, *Competition and Collective Bargaining in the Needle Trades: 1910–1967* (Ithaca, N.Y., 1972).

59. Graebner, "Coal-Mine Operator," p. 448.

60. Richard M. Abrams, *Conservatism in a Progressive Era: Massachusetts Politics, 1900–1912* (Cambridge, Mass., 1964), pp. ix, 20–22, 55, 71, 73, 77, 125–26.

61. Hackney, *Populism to Progressivism*, pp. 129, 131; Timberlake, *Prohibition and the Progressive Movement*, pp. 76–77.

Bibliographical Note

MANUSCRIPTS

Manuscript collections are especially important in constructing the national story of mine safety. The "Records of the United States Bureau of Mines," Record Group 70, at the National Records Center, Suitland, Maryland, fill some 500 boxes for the ten years after 1910. They are organized chronologically, by year, and within each year by subject, so that one may cull particular kinds of information quite easily. A comprehensive card index of correspondence in the collection is also available. Although the bureau was not established until 1910, the records contain some materials relevant to the workings of the Technologic Branch of the Geological Survey, the bureau's predecessor in mine safety. This is particularly important, since the records of the Technologic Branch were destroyed in a fire. There are scattered materials in Record Group 57, "Records of the Geologic Division of the United States Geological Survey," National Archives. The Manuscript Division of the Library of Congress maintains several collections of limited importance for this study. The papers of William Howard Taft and his secretary of the interior, Richard Achilles Ballinger (originals in the University of Washington Library, Seattle), were useful largely in interpreting the politics surrounding the appointment of Joseph A. Holmes as the first permanent bureau chief. The diaries of James R. Garfield, secretary of the interior under Theodore Roosevelt, and the papers of Gifford Pinchot, chief of the United States Forest Service and a close friend of Holmes, contain some information on the politics of safety, but the Pinchot papers were especially valuable for their insight into Holmes's character.

The best aggregation of manuscript materials on mine operators and safety exists in the West Virginia University Library, at Morgantown. Here are the papers of coal operators Justus Collins and Aretas Brooks Fleming (a former governor of the state), as well as the records of Collins's Winding Gulf Colliery Company. Both operators were active correspondents on a variety of issues, and Fleming played a crucial role in creating the Bureau of Mines. From the perspective of labor, the John Mitchell collection at the Mullen Library, Catholic University, Washington, D.C., was invaluable, in part because the files of the United Mine Workers of America, at UMWA Headquarters, Washington, D.C., were not open when I was researching (officials did allow me to use manuscript minutes of the National Executive Board Meetings which are printed but scattered). The State Letters of the Division of Mines, Department of Industrial Relations, State of Ohio, in the State Historical Society at Columbus, allow the only inside look (and hardly an extensive one) at the bailiwick of a chief mine inspector.

PERIODICALS

The dearth of materials for labor make the *United Mine Workers Journal* indispensable. Notable for its powerful editorials, the *Journal* follows major developments in the national and state politics of mine safety. The coal industry trade journals reflect regional interests but report and comment on events in every coalfield. Among the best are *Black Diamond* (for Illinois), *Coal and Coke Operator* and its successor *Coal and Coke Operator and Fuel Magazine, Coal Trade Bulletin,* and *Mines and Minerals. Coal Age* and *Colliery Engineer* are more oriented toward technology, engineering, and supervision. A number of these journals are difficult to find in full runs. Superior collections exist at the University of Illinois at Urbana, the Library of Congress, and the University of Pittsburgh.

GOVERNMENT PUBLICATIONS

The *Congressional Record* is important for understanding congressional opposition to the nationalization and bureaucratization of mine safety; otherwise, it is historical veneer. To get beneath that veneer, hearings and reports of commissions are essential. The most important were U.S., Congress, House, Industrial Commission, *Report of the Industrial Commission on Labor Legislation,* 19 vols. (1900–1902); U.S., Congress, House, Committee on Mines and Mining, *To Consider the Question of the Establishment of a Bureau of Mines: Hearings before the Committee on Mines and Mining,* 60th Cong., 1st sess., 1908; U.S., Congress, Senate, Immigration Commission, *Immigrants in Industries,* Part 1: *Bituminous Coal Mining,* 61st Cong., 2d sess., 1910, S. Doc. 633. The *Annual Reports* of the Bureau of Mines are basic to reconstructing the history of that agency, although for technical information the *Bulletins* of the bureau and the Geological Survey must be consulted.

Sources for state legislative history are frustratingly thin in content and a burden to use. One can trace the progress of legislation through House and Senate journals, but even this is laborious. There is no equivalent of the *Congressional Record* for the states I investigated (West Virginia's journals were the most informative). In short, the kind of interest group analysis that comes so easily at the national level is virtually impossible in the states. I found two special reports to be of value: Ohio, Legislature, *Report of the Ohio Coal Mining Commission to the Governor*, Part 2: *Prevention of Accidents*, 80th sess., 1913, House Joint Resolution 38, printed in Ohio, Senate, *Journal*, 1914, pp. 328–33; and West Virginia, Legislature, *Report of Hearings before the Joint Select Committee of the Legislature of West Virginia: To Investigate the Cause of Mine Explosions within the State and to Recommend Remedial Legislation Relating Thereto, Together with the Preliminary and Final Reports*, 1909, Substitute House Concurrent Resolution and House Joint Resolution 19. The *Bulletins* of the United States Department of Labor, Bureau of Labor Statistics, also follow state legislation. The *Reports* of the state mine inspectors fill in some of the gaps in the legislative picture and are vital for their insights into administration. They contain a wide variety of statistical information and are crucial for understanding the attitudes of operators, miners, and inspectors toward enforcement of the safety laws. The reports for West Virginia and Pennsylvania are the most revealing, those for Illinois (known as *Coal Reports*, and issued by the Bureau of Labor Statistics) the least.

OTHER PRIMARY MATERIALS

The published records of organizations are important for a number of topics. The American Mining Congress (AMC) *Proceedings* document the interplay between coal and metal mine operators and East and West and link coal-mining safety with conservation. The *Proceedings* of the National Safety Council (organized in 1912 as the First Cooperative Safety Congress) describe the relationship of safety to engineering and efficiency. The coal-dust controversy and other technical aspects of mine safety can be traced through the *Bulletins* and *Transactions* of the American Institute of Mining Engineers and the *Proceedings* of the Coal Mining Institute of America, an eastern regional body. The Illinois Coal Operators' Association published a *Monthly Bulletin* which I found useful for its records of miners and mine officials discharged for safety violations. The *Proceedings* of the Mine Inspectors Institute of America, founded in response to the 1907 Monongah disaster, reveal a factional organization in a continual state of disarray. With the AMC *Proceedings* and the Uniform Mining Laws Conference, 1916, *Report* (Springfield, Ill., n.d.), they are the basic source for my account of the movement for uniform state legislation.

The union perspective was developed from a variety of materials. The records of contract negotiations in Illinois (District 12 of the United Mine Workers of America [UMWA]), known as the Joint Conference or the Joint Convention, contain *verbatim* accounts of negotiations between operator and miner representatives on the safety issue. I found the *Proceedings* of UMWA conventions sketchy, though less so at the district than at the national level. The *Minutes of the International Executive Board* of the UMWA, while no less incomplete, are more intimate and more revealing of national labor priorities. Crystal Eastman's *Work-Accidents and the Law* (New York, 1910), one of the volumes in the *Pittsburgh Survey*, is a superb independent investigation.

SECONDARY LITERATURE

The historiography of the Progressive period is discussed in Chapter 6. Rather than summarize that analysis here, I shall focus on the literature that has some relevance to safety. There is no history of the safety movement, and one would be hard-pressed to put one together from the available studies, some of which are as much primary as secondary in nature. They include Lucian W. Chaney and Hugh S. Hanna, *The Safety Movement in the Iron and Steel Industry, 1907–1917*, U.S., Department of Labor, Bureau of Labor Statistics, Bulletin 234, no. 18: Industrial Accidents and Hygiene Series, June 1918 (Washington, D.C., 1918); E. H. Downey, *History of Work Accident Indemnity in Iowa* (Iowa City, 1912); Lucian W. Chaney, "Safety as Promoted by Federal Bureaus," *Proceedings of the National Safety Council* 2 (1913):101–7; Dianne Bennett and William Graebner, "Safety First: Slogan and Symbol of the Industrial Safety Movement," *Journal of the Illinois State Historical Society* 68 (June 1975):243–56; Edwin Higgins, "The Safety Movement in the Lake Superior Iron Region," *Bulletin of the American Institute of Mining Engineers* 94 (Oct. 1914):2515–30; Lew R. Palmer, "History of the Safety Movement," *Annals of the American Academy of Political and Social Science* 123 (Jan. 1926):9–19.

The history of mine safety in this period has been treated for certain geographical areas. Earl R. Beckner's *A History of Labor Legislation in Illinois* (Chicago, 1929) is outstanding. *The History of Legislation for the Protection of Coal Miners in Pennsylvania, 1824–1915* (New

York, 1942) by Alexander Trachtenberg is limited in imagination. The account in Katherine A. Harvey's *The Best-Dressed Miners: Life and Labor in the Maryland Coal Region, 1835–1910* (Ithaca, N.Y., 1969) is well done but does not place the subject in a larger context. The reader should also be aware of Glenn F. Massay, "Legislators, Lobbyists, and Loopholes: Coal Mining Legislation in West Virginia 1875–1901," *West Virginia History* 22 (April 1971):135–70, and K. Austin Kerr, "State Regulation of Coal Mines in the Nineteenth Century," Paper delivered at the Business History Conference, Northwestern University, Feb. 28, 1975, though they deal with an earlier period. Neither the United States Bureau of Mines nor the United Mine Workers has received adequate historical treatment. We also lack a thorough study of coal-industry economics for the Progressive years; I have summarized the assumptions about industrial structure which underpin the present study in "Great Expectations: The Search for Order in Bituminous Coal, 1890–1917," *Business History Review* 48 (Spring 1974):49–72.

Bibliography of Primary Sources

MANUSCRIPTS

American Association for Labor Legislation, Papers. Labor and Industrial Relations Institute, Cornell University, Ithaca, N.Y.

American Federation of Labor, Papers. Files, Office of the President, Samuel L. Gompers Correspondence, 1908, series 11, State Historical Society, Madison, Wisc.

Ballinger, Richard Achilles, Papers. Manuscript Division, United States Library of Congress, Washington, D.C., microfilm; originals in University of Washington Library, Seattle.

Bittner, Van Amberg, Papers. West Virginia University Library, Morgantown.

Collins, Justus, Papers. West Virginia University Library, Morgantown.

Dick, Charles, Papers. Ohio State Historical Society, Columbus.

Fleming, Aretas Brooks, Papers. West Virginia University Library, Morgantown.

Garfield, James R., Papers. Manuscript Division, United States Library of Congress, Washington, D.C.

Germer, Adolph, Papers. State Historical Society, Madison, Wisc.

Glasscock, William E., Papers. West Virginia University Library, Morgantown.

Holmes, Joseph Austin, Papers. Southern Historical Collection, University of North Carolina, Chapel Hill.

McDonald, Duncan, Papers. Illinois State Historical Society, Springfield.

Mitchell, John, Papers. Mullen Library, Catholic University, Washington, D.C.

Monongah Rescue Records. West Virginia University Library, Morgantown.

Ohio, Department of Industrial Relations, Division of Mines, State Letters. State Historical Society, Columbus.

Pinchot, Gifford, Papers. Manuscript Division, United States Library of Congress, Washington, D.C.

Taft, William Howard, Papers. Manuscript Division, United States Library of Congress, Washington, D.C.

United Mine Workers of America, Minutes, National Executive Board Meetings. United Mine Workers of America Headquarters, Washington, D.C.

United States Bureau of Mines, Records, Record Group 70. National Record Center, Suitland, Md.
United States Geological Survey, Records of the Geologic Division, Record Group 57. National Archives, Washington, D.C.
Walsh, Thomas J., Papers. Manuscript Division, United States Library of Congress, Washington, D.C.
Wilson, Woodrow, Papers. Manuscript Division, United States Library of Congress, Washington, D.C.
Winding Gulf Colliery Company, Papers. West Virginia University Library, Morgantown.

GOVERNMENT PUBLICATIONS

Illinois. Bureau of Labor Statistics. *Nineteenth Annual Coal Report of the Illinois Bureau of Labor Statistics, 1900.* Springfield, 1901. And reports for years 1904–1906, 1908–1910, 1919.
Kentucky. Inspector of Mines. *Annual Reports of the Inspector of Mines of the State of Kentucky, for 1901 and 1902.* C. J. Norwood, Chief Inspector. Prepared by the Chief Inspector. Louisville, 1903. And reports for 1903, 1904, 1907, 1908, 1914.
Ohio. *The Annotated Revised Statutes of the State of Ohio, Including All Laws of a General Nature in Force January, 1902*, ed. Clement Bates. 3d ed. 3 vols. Cincinnati, 1900.
Ohio. *The State of Ohio General and Local Acts Passed and Joint Resolutions Adopted by the 76th General Assembly, 1906.* Vol. 98. Springfield, 1906. And Volumes 99 (1908) and 100 (1910).
Ohio. Inspector of Mines. *Annual Report of the Chief Inspector of Mines to the Governor of the State of Ohio, for the Year Ending December 31, 1898.* Columbus, Ohio, 1899. And reports for 1899, 1905–1912, 1914.
Ohio. Legislature. *General and Local Acts Passed and Joint Resolutions Adopted by the General Assembly at Regular Sessions. Begun and Held in the City of Columbus, Session 1900.* Columbus, 1900. And miscellaneous years.
———. *Journal of the House of Representatives of the State of Ohio, for the Regular Session of the Seventy-fifth General Assembly, Commencing Monday, January 6, 1902.* Vol. 95. Cincinnati, 1902. And volumes for sessions beginning 1904, 1907, 1908, 1910, 1914.
———. *Journal of the Senate of the Seventy-sixth General Assembly of the State of Ohio for the Regular Session Commencing Monday, January 4, and Ending Monday, April 25, 1904.* Vol. 96. Springfield, 1904. And volumes for sessions beginning 1906, 1908, 1910 (two regular sessions).
———. *Report of the Ohio Coal Mining Commission to the Governor.* Part 2: *Prevention of Accidents*, printed in *Ohio Senate Journal, 1914,*

pp. 328–33. Dated December 17, 1913, under House Joint Resolution 38, 80th Session of the General Assembly. N.p., n.d.

Pennsylvania. *Digest of Pennsylvania Statute Law, 1920 (complete), based on Pepper and Lewis' Digest Laws of Pennsylvania.* St. Paul, Minn., 1921.

———. Department of Internal Affairs. Bureau of Mines. *Report of the Bureau of Mines of the Department of Internal Affairs of Pennsylvania, 1900.* N.p., 1901. And reports for 1901, 1903–1909, 1911, 1913, 1915.

———. Legislature. *Laws of the General Assembly of the Commonwealth of Pennsylvania, Session 1899.* Harrisburg, 1899. And other (odd) years.

———. *Legislative Journal for the Session of 1911.* Harrisburg, 1912.

U.S. Bureau of the Census. *Historical Statistics of the United States, Colonial Times to 1957.* Washington, D.C., 1960.

U.S. Coal Commission. *Report of the United States Coal Commission Transmitted pursuant to the Act Approved September 22, 1922.* Vol. 3 of 5 vols. Sen. Doc. 195, 68th Cong., 2d sess. Washington, D.C., 1925.

U.S. Committee on Public Information. "Studying Industrial Safety Code with View to Securing Uniformity." *Bulletin*, vol. 3, no. 52 (Jan. 29, 1919):9.

U.S. Congress. *Congressional Record.* Miscellaneous years.

U.S. Congress. House. *Estimate for Investigation of Mine Accidents, etc., by Geological Survey.* H. Doc. 523, 60th Cong., 1st sess., vol. 6, ser. 5375, 1908.

———. *Permissible Explosives Tested prior to January 1, 1911, and Precautions to Be Taken in Their Use.* By Clarence Hall, Bureau of Mines, H. Doc. 1405, 61st Cong., 2d sess., ser. 6069, 1911.

———. *Protection of Lives of Miners in the Territories.* H. Rept. 148, 57th Cong., 1st sess., vol. 1, ser. 4399, 1902.

———. *Report on the Miners' Strike in Bituminous Coal Field in Westmoreland County, Pa., in 1910–11.* H. Doc. 847, 62d Cong., 2d sess., ser. 6279, 1912.

———. Committee on Mines and Mining. *Appropriations for Mining Schools*, Hearings before the Committee on Mines and Mining, House of Representatives, on H.R. 6063, 63d Cong., 2d sess., 1913.

———. Committee on Mines and Mining. *Bill to Change Law Establishing Bureau of Mines*, Hearings before the Committee on Mines and Mining, House of Representatives, 1912.

———. Committee on Mines and Mining. *Bureau of Mines and Mining.* H. Rept. 1065, 47th Cong., 1st sess., ser. 2068, 1882.

———. Committee on Mines and Mining. *Mining Experiment Station in Colorado*, Hearings before the Committee on Mines and Mining, House of Representatives, on H.R. 11414, 1911.

———. Committee on Mines and Mining. *To Consider the Question of the Establishment of a Bureau of Mines*, Hearings before the Committee on Mines and Mining, House of Representatives, 1908.

———. Industrial Commission. *Report of the Industrial Commission on Labor Legislation.* Vol. 5 of 19 vols. (1900–1902). Washington, D.C., 1900.

———. Industrial Commission. *Report of the Industrial Commission on the Relations and Conditions of Capital and Labor Employed in the Mining Industry.* H. Doc. 181, 57th Cong., 1st sess., ser. 4342. Vol. 12 of 19 vols. (1900–1902). Washington, D.C., 1901.

U.S. Congress. Senate. *Estimates for Urgent Deficiency, Bureau of Mines.* S. Doc. 157, 62d Cong., 2d sess., ser. 6180, 1912.

———. *A Report on Labor Disturbances in the State of Colorado, from 1880 to 1904, Inclusive.* S. Doc. 122, 58th Cong., 3d sess., vol. 3, ser. 4765, 1905.

———. Anthracite Coal Strike Commission. *Report to the President on the Anthracite Coal Strike of May–October, 1902, by the Anthracite Coal Strike Commission.* S. Doc. 6, 58th Cong., special sess., vol. 1, ser. 4556, 1903.

———. Commission on Industrial Relations. *Industrial Relations.* Final Report and Testimony, 11 vols. Vol. 8, S. Doc. 405, 64th Cong., 1st sess., ser. 6935, 1916.

———. Committee on Mines and Mining. *Bureau of Mines.* S. Rept. 353, to accompany H.R. 13915, from Committee on Mines and Mining, Senate, 61st Cong., 2d sess., ser. 5583, 1910.

———. Committee on Mines and Mining. *Division of Mines and Mining, United States Geological Survey.* S. Doc. 40, 55th Cong., 3d sess., 1898.

———. Committee on Mines and Mining. *Establishing Bureau of Mines in Interior Department.* S. Rept. 692, to accompany H.R. 20883, 60th Cong., 1st sess., vol. 2, ser. 5219, 1908.

———. Committee on Public Health and National Quarantine. *Proposed Department of Public Health,* Hearings before the Committee on Public Health and National Quarantine, Senate, on S. 6049, 1910.

———. Committee of the Senate upon the Relations between Labor and Capital. *Report of the Committee of the Senate upon the Relations between Labor and Capital, and Testimony Taken by the Committee.* 48th Cong., 5 vols., 1: *Testimony,* 1885.

———. Immigration Commission. *Immigrants in Industries.* Part 1: *Bituminous Coal Mining.* 25 vols. S. Doc. 633, 61st Cong., 2d sess., ser. 5667, 1910.

U.S., Department of Commerce. Bureau of the Census. *Thirteenth Census of the United States,* Volume 11: *Mines and Quarries (1909), General Report and Analysis.* Washington, D.C., 1913.

U.S., Department of Commerce and Labor. Bureau of Labor. *Bulletin* 52. Washington, D.C., 1904.

———. Bureau of Labor. *Bulletin* 54. Vol. 9. Washington, D.C., 1904.

U.S., Department of the Interior. Bureau of Mines. *Achievements in Mine*

Safety Research and Problems Yet to Be Solved. By Arno C. Fieldner, Information Circular 7573, June 1950.

———. Bureau of Mines. *First Annual Report of the Director of the Bureau of Mines to the Secretary of the Interior, for the Fiscal Year Ended June 30, 1911.* Joseph A. Holmes, director. Washington, D.C., 1912. And years 1912–1919.

———. Bureau of Mines. *The Joseph A. Holmes Safety Association and Its Awards.* By D. Harrington, Louise Pedlow, and Anna P. Brown. Bulletin 421. Washington, D.C., 1940.

———. Bureau of Mines. *First National Mine-Safety Demonstration, Pittsburgh, Pennsylvania, October 30 and 31, 1911.* By Herbert M. Wilson and Albert H. Fay, with a chapter on the explosion at the experimental mine by George S. Rice. Bulletin 44. Washington, D.C., 1912.

———. Bureau of Mines. *Safety in the Mining Industry.* By D. Harrington, J. H. East, Jr., and R. G. Warncke. Bulletin 481. Washington, D.C., 1950.

———. Bureau of Mines. Office of Minerals Reports. Joseph A. Holmes sketch. (Typewritten.)

———. Geological Survey. *Coal-Mine Accidents: Their Causes and Prevention: A Preliminary Statistical Report.* By Clarence Hall and Walter O. Snelling, introduction by Joseph A. Holmes, Technologic Branch. Bulletin 333. Washington, D.C., 1907.

———. Geological Survey. *Mineral Resources of the United States, 1900.* Washington, 1901. And years 1901–1920.

———. Geological Survey. *The Prevention of Mine Explosions, Report and Recommendations.* By Victor Watteyne, Carl Meissner, and Arthur Desborough. Bulletin 369. Washington, D.C., 1908.

U.S., Department of Labor. Bureau of Labor Statistics. *Labor Laws of the United States, with Decisions of Courts Relating Thereto.* Bulletin 148, 2 parts, 1914.

West Virginia. *The Code of West Virginia,* 1906, National Reporter System. St. Paul, Minn., 1906.

———. *Hogg's West Virginia Code, Annotated,* ed. Charles E. Hogg. 3 vols. St. Paul, Minn., 1914.

———. *West Virginia Public Documents, 1917–1918.* 3 vols. N.p., n.d.

———. Inspector of Mines. *Sixteenth, Seventeenth and Eighteenth Annual Reports. Coal Mines in the State of West Virginia, U.S.A. For the Years Ending June 30, 1898, 1899, 1900.* By James W. Paul, Chief Mine Inspector. Charleston, n.d. And reports for years 1903–1910, 1912, 1914–1915.

———. Legislature. *Acts of the Legislature of West Virginia, Session 1899.* Charleston, 1899. And other (odd) years.

———. *Journal of the House of Delegates of the State of West Virginia for the Twenty-sixth Regular Session Commencing January 14, 1903.*

Charleston, 1903. And volumes for sessions beginning 1905, 1907, 1908 (extra session).
———. *Journal of the Senate of the State of West Virginia for the Twenty-fifth Regular Session. Commencing January 9, 1901.* Charleston, 1901. And volumes for sessions beginning 1907, 1908 (and extra session), 1909, 1915, 1917.
———. *Report of Hearings before the Joint Select Committee of the Legislature of West Virginia, Appointed under Substitute for House Concurrent Resolution No. 5 and House Joint Resolution No. 19, to Investigate the Cause of Mine Explosions within the State and to Recommend Remedial Legislation Relating Thereto, together with the Preliminary and Final Reports.* Charleston, 1909.

GENERAL WORKS

American Bar Association. *Journal.* Vol. 2 (1916).
American Institute of Mining Engineers. *Bulletin* 74 (Feb. 1913):319–29.
American Institute of Mining Engineers. *Transactions* 13 (Feb. 1884–June 1885). And years 1881–1882, 1894, 1909, 1910, 1914 and 1917.
American Mine Safety Association. *First Annual Transactions, Constitution and By-Laws.* Held at Pittsburgh, Pennsylvania, January, 1914. N.p., n.d. And 1915.
American Mining Congress. *Proceedings of the Sixth Annual Session.* Held at Deadwood and Lead, South Dakota, September 7–12, 1903. [Vol. 6.] Portland, Ore., 1904. And 1904–1919.
Andrews, John B. "Needless Hazards in the Coal Industry." *Annals of the American Academy of Political and Social Science* 111 (Jan. 1924):24–31.
Andros, S. D. "Coal Mining Practice in District IV." Illinois Coal Mining Investigations Co-operative Agreement, *Bulletin* 12. Urbana, Ill., 1915.
Boyd, James H. *Workmen's Compensation.* 2 vols. Indianapolis, 1913.
Bradshaw, George. *Safety First.* New York, 1913.
Chaney, Lucian W., and Hanna, Hugh S. *The Safety Movement in the Iron and Steel Industry, 1907–1917.* U.S., Department of Labor, Bureau of Labor Statistics, *Bulletin* 234, no. 18. Industrial Accidents and Hygiene Series, June 1918 (Washington, D.C., 1918).
Claghorn, Kate Holladay. *The Immigrant's Day in Court.* New York, 1923.
Coal Mining Institute of America. *Proceedings,* 1903–1904–1905. Pittsburgh, 1906.
Coal Operators of West Virginia and Other States. *Proceedings of Meeting, January 8, 1908.* Washington, D.C. N.p., n.d.
Dague, J. H., and Phillips, S. J. *Mine Accidents and Their Prevention.* New York, 1912.
Devine, Edward T. *Coal.* Bloomington, Ill., 1925.
Downey, E. H. *History of Work Accident Indemnity in Iowa.* Iowa Eco-

nomic History Series. Edited by Benjamin F. Shambaugh. Iowa City, 1912.

Earp, Edwin L. *The Social Engineer.* New York, 1911.

Eastman, Crystal. *Work-Accidents and the Law.* Vol. 1 of *Pittsburgh Survey*, 6 vols. New York, 1910.

Evans, Chris. *History of the United Mine Workers of America.* 2 vols. Indianapolis, n.d.

Fairmont Coal Company. *The Explosion at Monongah Mines, Fairmont Coal Company. Bulletin* 11 by Frank Haas, supplementing *Annual Report of Operation*, 1907. Fairmont, W. Va., 1908.

Fay, Albert H. "Mining Accidents and Uniform Records." *Proceedings of the Second Pan American Scientific Congress* 8 (1915–1916):494–515.

Frick Coke Company. *Safety the First Consideration.* N.p., 1916.

Gompers, Samuel. "Strikes and the Coal-Miners." *Forum* 24 (Sept. 1897): 27–33.

Goodrich, Carter. *The Miner's Freedom.* Boston, 1925.

Hambridge, Jay. "An Artist's Impressions of the Colliery Region." *Century Magazine* 55 (April 1898):822–28.

Hamilton, Walton H., and Wright, Helen R. *The Case of Bituminous Coal.* New York, 1925.

Hapgood, Powers. *In Non-Union Mines: The Diary of a Coal Digger in Central Pennsylvania, August–September, 1921.* New York, n.d.

Holmes, Joseph A. "Government Measures to Increase Mine Safety." *Annals of the American Academy of Political and Social Science* 38 (1911): 112–14.

Illinois Coal Operators. *Monthly Bulletin.* Vol. 1, no. 1 (Dec. 1908). N.p., n.d.

Industrial Workers of the World. *Coal-Mine Workers and Their Industry.* Prepared by the Education Bureau of the IWW for Coal-Mine Workers Industrial Union 220, IWW. N.p., n.d.

International Commerce. *Coal Trade of the U.S., 1900.* N.p., 1900.

International Mining Congress. *Transactions* (1915).

Interstate Convention of 1908. Joint Conference of Coal Miners and Operators of Western Pennsylvania, Ohio and Indiana. Held at Toledo, Ohio, April 14–17, 1908. N.p., n.d.

Interstate Conventions, 1906. Joint Conferences of Coal Miners and Operators of Western Pennsylvania, Ohio, Indiana, and Illinois. Indianapolis, Indiana, Jan. 25–Feb. 2, March 20–29, 1906. N.p., n.d.

Interstate Joint Conference. Proceedings, Coal Operators and Coal Miners of Western Pennsylvania, Ohio, Indiana and Illinois. Indianapolis, Indiana, Jan. 26 to Feb. 1, 1912, Cleveland, Ohio, March 20 to 30, 1912. N.p., n.d.

Joint Conference of the Illinois Coal Operators Association, the Coal Operators Association of the Fifth and Ninth Districts of Illinois, the Central

Illinois Coal Operators Association and the UMWA *District No. 12.* Peoria, Illinois, April 2 to May 8, 1914. N.p., n.d.
Joint Convention of the Illinois Coal Operators Association and the UMWA *District 12.* Held at Peoria, Illinois, Feb. 24 to March 13, 1902. N.p., n.d. Also 1903, 1904, and 1906.
Jones, D. R. *The Mining Conflict.* Pittsburgh, 1880.
Jones, [Mary Harris]. *Autobiography of Mother Jones.* Edited by Mary Field Parton. Chicago, 1925.
Lane, Winthrop D. *Civil War in West Virginia.* New York, 1921.
Manning, Van H. "Mine Safety Devices Developed by the United States Bureau of Mines." *Annual Report of the Board of Regents of the Smithsonian Institution, for the Year Ending June 30, 1916.* Washington, D.C., 1917, pp. 533–44.
———. "The United States Bureau of Mines." *Proceedings of the Second Pan American Scientific Congress* 8 (1915–1916):484–92.
Mine Inspectors Institute of the United States of America. *Proceedings.* Held at Indianapolis, June 1908. N.p., n.d. And years 1909–1910, 1912–1915, 1919, 1920.
Mining and Metallurgical Society of America. *Bulletin* 42, no. 4 (Dec. 1911).
Mitchell, Guy Elliot. "The New Bureau of Mines." *World To-Day* 29 (July–Dec. 1910):1150–55.
Mitchell, John. *Organized Labor.* Philadelphia, 1903.
———. *The Wage Earner and His Problems.* Washington, D.C., 1913.
Moran, E. L. "The Coal Traffic of the Great Lakes." *Journal of Geography* 15 (Jan. 1917):150–59.
National Fire Protection Association. *Newsletter.* Nos. 2, 4, 5, 6–16, 17, 18–30, 33–39, 41, 42. Boston, 1917–1920.
———. *Proceedings of the Fifteenth Annual Meeting.* Held at New York, May 23–25, 1911. N.p., n.d. And 1912.
National Safety Council. *Proceedings of the First Cooperative Safety Congress.* Held at Milwaukee, Wisconsin, September 30–October 5, 1912. [Vol. 1.] N.p., n.d. [This first congress was held under the auspices of the Association of Iron and Steel Electrical Engineers.] And years 1915–1919.
Palmer, Lew R. "History of the Safety Movement." *Annals of the American Academy of Political and Social Science* 123 (Jan. 1926):9–19.
Powell, Fred Wilbur. *The Bureau of Mines.* No. 3. Service Monographs of the United States Government, Institute for Government Research, New York, 1922.
Roberts, Peter. *Anthracite Coal Communities.* New York, 1904.
Rogers, Jack. "I Remember That Mining Town." *West Virginia Review* 15 (April 1938):203–5.
Ross, David. *History of Coal Mining in Illinois.* An Address Delivered be-

fore the State Mining Institute in Annual Session at DuQuoin, Illinois, May 26th, 1916. Springfield, Ill., n.d.

Roy, Andrew. *A History of the Coal Miners of the United States.* Columbus, Ohio, n.d.

"Safety First in Mining." *Pan American Union Bulletin* 49 (Oct. 1919): 428–36.

Saliers, Earl A. *The Coal Miner.* N.p., 1912.

Shockley, William H. "The Economic and Social Influence of Mining with Special Reference to the United States." *Transactions of the International Engineering Congress* (1915), pp. 1–66.

Simons, A. M. *Wasting Human Life.* N.p., n.d. [Probably Chicago, ca. 1911.]

Tolman, William H., and Kendall, Leonard B. *Safety.* New York, 1913.

Uniform Mining Laws Conference. *Report.* Held at Chicago, Illinois, November 13, 14, 15, 1916. Springfield, n.d.

United Mine Workers of America. *Minutes of the Thirteenth Annual Convention.* Held at Indianapolis, January 20–29, 1902. Indianapolis, 1902. And years 1905, 1906, 1908–1910, 1912, 1914, 1916, 1918.

———. District 2. *Minutes of the Nineteenth Annual Convention of District Number 2*, DuBois, Pennsylvania, March 24–30, 1908. N.p., n.d. And 1911 (22nd).

———. District 12. *Proceedings of the Twenty-first Annual Convention of the United Mine Workers of America, District 12.* Held in the City of Peoria, Illinois, Feb. 15th to March 5th, Inclusive, 1910. N.p., n.d. And years 1910 (reconvened 21st convention), 1911 (22nd), and 1913 (24th).

———. International Executive Board. *Minutes of the International Execcutive Board, March 30, 1908 Meeting.* Indianapolis, Indiana. N.p., n.d.

Van Hise, Charles Richard. *The Conservation of Natural Resources.* New York, 1910.

Virtue, G. O. "The Anthracite Mine Laborers." U.S., Department of Labor, *Bulletin* 13 (Nov. 1897):728–74.

Warne, Frank Julian. "The Effect of Unionism upon the Mine Worker." *Annals of the American Academy of Political and Social Science* 21 (Jan.–June 1903):20–35.

Zillmer, Raymond Theodore. "The Commissioners on Uniform State Laws." *American Law Review* 47 (Nov.–Dec. 1913):864–90.

Index